The NEA Committee on the Safety of Nuclear Installations (CSNI) is an international committee made up of scientists and engineers who have responsibilities for nuclear safety research and nuclear licensing. The Committee was set up in 1973 to develop and co-ordinate the Nuclear Energy Agency's work in nuclear safety matters, replacing the former Committee on Reactor Safety Technology (CREST) with its more limited scope.

The Committee's purpose is to foster international co-operation in nuclear safety amongst the OECD Member countries. This is done essentially by :

i) exchanging information about progress in safety research and regulatory matters in the different countries, and maintaining banks of specific data; these arrangements are of immediate benefit to the countries concerned;

ii) setting up working groups or task forces and arranging specialist meetings, in order to implement co-operation on specific subjects, and establishing international projects; the output of the study groups and meetings goes to enrich the data base available to national regulatory authorities and to the scientific community at large. If it reveals substantial gaps in knowledge or differences between national practices, the Committee may recommend that a unified approach be adopted to the problems involved. The aim here is to minimise differences and to achieve an international consensus wherever possible.

The main CSNI activities cover particular aspects of safety research relative to water reactors and fast reactors; probabilistic assessment and reliability analysis, especially with regard to rare events ; siting research ; fuel cycle safety research ; various safety aspects of steel components in nuclear installations ; and a number of specific exchanges of information.

The Committee has set up a sub-Committee on Licensing which examines a variety of nuclear regulatory problems, provides a forum for the free discussion of licensing questions and reviews the regulatory impact of the conclusions reached by CSNI.

At its sixth meeting in November 1978, the CSNI decided to sponsor the preparation of state-of-the-art reports on selected subjects related to nuclear safety research, in order to review the existing state of knowledge, identify possible gaps and recommend additional research work when necessary. Three reports on Nuclear Aerosols in Reactor Safety, Reference Seismic Ground Motions in Nuclear Safety Assessments and Fuel Cycle Safety were published in 1979,1980 and 1981 respectively in the CSNI "State-of-the-Art-Report" series. The present report is the fourth in this series.

The mathematical modelling of the critical flow that occurs during a Loss of Coolant Accident is one of the safety topics listed for priority consideration by the Nuclear Energy Agency Committee on the Safety of Nuclear Installations (CSNI) in the area of water reactor safety.

During the November 1978 meeting of the CSNI, the Committee decided that it would be useful to prepare state-of-the-art reports "SOARS" on the safety questions that it was dealing with. These reports would summarise the current level of understanding on a given topic and would point out what gaps in this understanding still remained. Further, they would include conclusions on the state-of-the-art as well as recommendations for future work needed to improve understanding in a given area.

It was decided that these reports would not attempt to be comprehensive technical documents, but would be critical surveys of the given area to provide technical managers assistance in the decision-making process. At the same time, the SOAR would provide brief discussion and references to the latest work in the field for use at the technical or research level.

At its November 1979 meeting, the CSNI considered the priority list of topics for SOARs that had been drawn up from the suggestions received by the Secretariat from the CSNI and selected critical flow modelling for treatment in one of the first SOARs in the water reactor safety research area.

A meeting of the experts able to contribute to this SOAR was held in January 1980 to outline the contents and select experts to draft individual chapters. This group met again in May 1980 to exchange and review these first drafts. Following overall editing, the report was reviewed by members of the CSNI Working Group on Emergency Core Cooling and Fuel Behaviour in Water Reactors.

Members of the group of experts who participated in the production of this report were:

Mr. Ian Brittain (SOAR Editor, drafted Chapter 6)	Atomic Energy Establishment, Winfrith United Kingdom Atomic Energy Authority, Dorchester, United Kingdom
Prof. Dr-Ing H. Karwat (Chapters 1 and 2)	Technische Universität München, Munich, Federal Republic of Germany
Mr. F. D'Auria (Chapter 3)	Instituto di Impianti Nucleari, University of Pisa, Pisa, Italy
Mr. P. Vigni (collaborated on Chapter 3)	Instituto di Impianti Nucleari, University of Pisa, Pisa, Italy
Mr. D. Hall (Chapter 4)	EG&G Idaho Incorporated, Idaho Falls, Idaho, USA

Dr. M. Réocreux (Chapter 5) — Départment de Sûreté Nucléaire, Commissariat à l'Energie Nucléaire, Fontenay-aux-Roses, France

In Chapter 3 of this SOAR, it was intended to provide a survey of the available critical flow models. The work of Messrs D'Auria and Vigni, sponsored by the Comitato Nazionale per l'Energia Nucleare, resulted in descriptions of over sixty such models. Since this material was too voluminous to meet the objectives of the SOAR, only those models used in RELAP4 have been treated herein. The larger comprehensive survey of critical flow models has been published as a technical addendum to this report (CSNI Report No. 49).

TABLE OF CONTENTS

		Page
1.	INTRODUCTION	10
2.	REQUIRED ACCURACY FOR CRITICAL FLOW MODELS	11
	2.1 The Blowdown Process	11
	2.2 Understanding Experimental Results	12
	2.3 Design of Structures	13
3.	SURVEY OF AVAILABLE CRITICAL FLOW MODELS	15
	3.1 Main Aspects of Critical Flow	15
	3.2 Difficulties in Analysing Two-Phase Flow	16
	3.3 Classification of the Models	17
	3.3.1 Criteria	17
	3.3.2 List of Models by Category	19
	3.4 Outline of the Analysis	25
	3.4.1 General	25
	3.4.2 Discussion of the Main Assumptions ...	25
	3.5 Description of Models in RELAP4	28
	3.5.1 Sonic Model	28
	3.5.2 MOODY Model	28
	3.5.3 HENRY-FAUSKE Model	29
	3.5.4 Homogeneous Equilibrium Model	30
	3.5.5 Discussion	30
	3.6 Comparison of the Models	31
	3.6.1 General	31
	3.6.2 Quantitative Comparison	32

		Page
3.7	Conclusions on Available Models	33
	3.7.1 Main Conclusions of Some Authors	33
	3.7.2 Results and Discussion	36
	List of Symbols Used	38
	Figures	39-48
	References	49
4.	CRITICAL FLOW EXPERIMENTAL DATA	50
4.1	Data Base Inventory	50
4.2	Discussion of the Data Base	59
4.3	Conclusions and Recommendations	60
	4.3.1 Conclusions	60
	4.3.2 Recommendations	61
	Figures	63-65
	References	66
5.	APPLICATIONS OF CRITICAL FLOW MODELS IN SAFETY CALCULATIONS	70
5.1	Introductory Remarks	70
5.2	List of Models used in Safety Calculations	70
	5.2.1 RELAP4 Code	71
	5.2.2 RELAP5 Code	74
	5.2.3 TRAC Code	76
	5.2.4 DRUFAN Code	77
	5.2.5 CATHARE Code	78
5.3	Discussion	79
	5.3.1 Discussion and Comparison of Physical Models	79
	5.3.2 Discussion and Comparison of Numerical Applications	80
	5.3.3 Discussion of Main Trends Shown by the Results	84

Page

5.4 Recommendations 87

5.4.1 Practical Calculations: Test and Safety Calculations 87

5.4.2 Physical Modelling 87

5.4.3 Numerical Application 88

5.4.4 Confirmatory Testing of Models 89

5.5 Conclusions 89

List of Symbols Used 91

Figures 93

References 94

Appendix 5.1 Example of the Use of Stagnation Conditions in Single-Phase Flow ... 95

Appendix 5.2 Summary of the Physical Analysis of Mechanisms Governing Single- and Two-Phase Critical Flow 96

6. GENERAL CONCLUSIONS AND RECOMMENDATIONS 101

6.1 Conclusions 101

6.2 Recommendations 102

1

INTRODUCTION

In the case of a loss-of-coolant accident (LOCA) in a water-cooled nuclear power reactor, coolant is lost through circuit ruptures or inadvertantly opened valves.

As long as the system pressure is larger than a certain critical pressure compared with the environment to which the coolant is released, sound velocity limits the mass discharge rate through the phenomenon of critical flow. Flow area and critical mass flow rates determine the energy and mass loss of the primary system. The limiting sound velocity is strongly dependent on the property of the discharged fluid, particularly if a two-phase steam-liquid mixture is discharged. For single-phase gas or steam flow, the prediction of the relevant sound velocity of two-phase mixtures and the associated mass discharge rates are the subject of a number of worldwide research activities in the area of reactor safety as well as space propulsion research.

Of particular relevance to reactor safety is the study of critical mass flow behaviour for steam-liquid mixtures. The analytical description and prediction of critical discharge rates plays an important role in the design of engineered safeguards systems for large nuclear power reactors and in the interpretation of associated experiments. Experimental verification of analytical models for critical mass flow, on the other hand, is highly dependent on the experimental situation and the available measurement techniques.

The purpose of this report is to describe the technical and scientific background which is available to understand the phenomenon "Critical Two-Phase Flow" and to what extent it is amenable to analytical description by or within computer models. The report is based on applicable information available today and does not go into detail about the historic development of knowledge of critical two-phase flow prediction and determination.

In order to place in perspective the importance of critical flow modelling, Chapter 2 discusses the relevant safety questions for which a sound knowledge of the critical mass flow rate phenomenon is necessary. Chapter 3 provides a survey of the available critical flow models and Chapter 4 discusses the experimental data base that exists to support these models. Chapter 5 presents a summary of the models and modelling techniques used in the so-called advanced (non-equilibrium) codes that are now being developed and used for reactor safety analyses. Chapter 6 presents the overall conclusions of the report and sets out the recommendations for improving the state-of-the-art of critical flow modelling.

2

REQUIRED ACCURACY FOR CRITICAL FLOW MODELS

The discharge of hot water from the coolant system of water-cooled power reactors, should a loss-of-coolant accident occur, has an important influence on the behaviour of the coolant system and associated engineered safeguards designed to cope with such emergency situations. The absolute magnitude of the flow and its variation with time are both important parameters. However, we can distinguish between those design items which are directly dependent on an accurate knowledge of the magnitude of the discharge rates and other design features where the coolant discharge rate is only one important parameter out of several having direct impact on accident considerations. A third area where an accurate knowledge of the coolant discharge is directly required covers fluid dynamics model experiments, where the simulated loss-of-coolant is a very important experimental boundary condition in interpreting the results of small scale experiments and their extrapolation to full scale units.

The first area mentioned above involves mainly the mechanical effects of a spreading depressurisation wave on internals within the primary coolant system and the corresponding effects of a rapid pressure build-up within the surrounding containment system. The second area of interest is the blowdown process itself and the interrelation with emergency core cooling systems. Here, the influence of a wide range of parameters has to be investigated to demonstrate efficient emergency core cooling.

Taking a full understanding of loss-of-coolant experiments as the third area of interest, mass and energy discharge should be measured accurately, but these measurements are sometimes difficult and have only limited accuracy. Therefore, it is often desirable to back-up such measurements with an analytical determination of critical mass flow.

Frequently, experiments, in particular integral loop experiments, also serve the purpose of code verification. The character of these experiments is largely determined by the mass and energy discharge (large break/small break-simulations) making accurate determination of these two factors mandatory.

2.1 THE BLOWDOWN PROCESS

To predict the efficiency of emergency core cooling systems for water-cooled power reactors, the relevant rules and guidelines require that the whole spectrum of possible break locations and break sizes be investigated to determine the most unfavourable situations with which emergency core cooling systems have to cope. The cooling conditions of the core are dominated by the coolant flow distribution throughout the whole primary system. Location and duration of possible flow stagnation within the core are governed by the location and size of a rupture, the operating conditions of the recirculation pumps, injection of emergency coolant, and the mass and energy discharge through the leak. The absolute magnitude of the leakage flow in a particular situation, however, is of secondary importance for questions raised in licensing procedures, for inaccuracies in this parameter might be covered by variation of the two other important parameters, "size" and "location" of the rupture. For example, an error of 20% in the determination of the absolute mass discharge might be covered by a variation in the break area of about 20%.

However, the relative magnitude of the mass loss is important in terms of its time variation. In particular, for the predictive analysis of the consequences of small leaks the time variation of mass and energy discharge determines the formation of liquid levels and the starting times at which emergency core cooling systems become operative. This is particularly true if small or medium sized breaks are assumed to occur through a failure of pipes connected with the main coolant system. These connecting lines are clearly defined with respect to size and location and less subject to fracture mechanics speculations. On the other hand, small or medium sized LOCAs result in slow coolant flow transients, while stored energy from fuel may be extracted and clad temperatures are reduced during the blowdown period. Therefore, small errors in the timewise prediction of the transient critical two-phase flow rates are of minor importance.

In general, the requirements on accuracy in predicting critical mass flow rates during the blowdown period with respect to its effect on emergency core cooling are not stringent when applied to the accident analyses of power reactor systems.

2.2 UNDERSTANDING EXPERIMENTAL RESULTS

A considerable number of experiments are carried out to study fundamental questions on processes involved in the reactor safety analyses. On the one hand, these experiments serve to provide fundamental knowledge. Simultaneously they serve as a basis for the experimental verification of computer codes which are in use for the design of reactor systems. There are also integral experiments carried out to understand systems behaviour, and the attempt is made to extrapolate experimental observations directly to full-sized reactor systems, with or without proper application of modelling laws.

These goals require an accurate knowledge of the technical conditions under which the experiment has been carried out. The simulation of a rupture and the resulting critical mass flow through the rupture is one of the most important aspects of fluid dynamic experiments to investigate emergency core cooling processes or the behaviour of the various types of containment systems.

To understand completely the transient results of experiments in terms of pressures, temperatures and measured densities requires the best possible knowledge about the operating conditions such as heat transfer from structures, pump behaviour for two-phase conditions, behaviour of valves, and "discharge of liquid and energy at the rupture". Otherwise the danger of misinterpretation of an experimental observation in one or other respects cannot be excluded.

Observed disagreement between experimental and analytical predictions of the pressure or temperature transients are quite often caused by a poor prediction of the mass and energy discharge rates at the location of the simulated rupture. Therefore, it is sometimes helpful to use in those predictive calculations an "as measured" input function for these variables to avoid possible misinterpretation of the comparison of other measured and analytically predicted parameters. This, of course, is not relevant to those cases where determination of the mass-discharge rates is the main objective of the experiment.

Where mass flow rates and associated specific enthalpies are important parameters, the question of accuracy of measurements arises. In recent years several research activities have been undertaken in order to improve measurement technique for both mass flow and critical mass flow in particular.

The results of these efforts, however, have been limited. Therefore, attempts are made to combine measurements and analyses to obtain a more precise determination of experimentally observed mass and energy discharge in test facilities. In particular better measurement accuracy is required for the most interesting initial phase of mass discharge experiments.

2.3 DESIGN OF STRUCTURES

The philosophy of the Design Basis Accident requires that water-cooled power reactors have to be analysed for the consequences of the rupture of a primary coolant line up to the maximum installed size connected to the reactor pressure vessel. It is evident that the maximum possible effects of depressurisation would be expected if large ruptures occur, and specifically if such large ruptures occur in the vicinity of important structures.

Important structures within the primary coolant system are in particular the reactor pressure internals with the core support structure, the control rods and their guide tubes, the fuel elements themselves and the core barrel. The behaviour of check valves is also of interest for boiling water reactors for anticipated ruptures of the feed water lines.

Within the containment system the loading of structures is directly dependent on the size, location and discharge rate of possible ruptures. The design of pressure suppression systems used for boiling water reactors likewise depends on a knowledge of the maximum possible discharge rates which have to be taken into consideration within the Design Basis Accident philosophy.

For a given design concept the most unfavourable location and size of an anticipated rupture can easily be defined. In order to determine the depressurisation effects, two other essential parameters remain open. These are the formation of the fracture (fracture mechanics) and the determination of the leakage flow. To cope with the uncertainty of fracture mechanics predictions a nearly instantaneous formation of a rupture is normally assumed.

In particular, for situations where pressurised water is involved, coupled fluid dynamics and structural mechanics codes (based on non-equilibrium considerations) are in use to determine loading on neighbouring structures, discharge rates at the location of the rupture being the most important factor influencing the calculated results.

Concerning the short term development of leakage flow through the assumed rupture, the size of the limiting critical mass-flow becomes a direct input to predictive analyses.

Some codes require the selection of options for the assumed critical mass flow correlation, either in form of tables or by calling for empirical correlations dependent on upstream conditions. Some codes calculate critical mass flow rates by calculating choking conditions (Mach No. = 1) automatically. The latter group of codes, however, requires the specification of parameters governing thermodynamic and/or mechanical non-equilibrium between the steam and liquid phases.

If accurate knowledge of the loading conditions is required a best estimate description of critical mass flow rates is necessary. On the other hand, the interpretation of experiments and the comparison between experiment and theory require the best possible knowledge about critical mass flow rates, nevertheless some uncertainty bands will remain. Insufficient knowledge about critical mass flow rates during the experiment frequently is one reason for disagreement between experiment and theory.

Safety factors which have to be added to the results of best estimate calculations for expected loadings in real reactor systems have to be chosen according to this uncertainty band. Sometimes we still find so-called conservative approaches which originate in large part from the approximate treatment of the sub-cooled sonic decompression wave and its loading effects on important structures. Such methods completely neglect the formation of voids during the initial phase of depressurisation and are not applicable to a more advanced treatment of non-equilibrium processes in two-phase flow.

Similar observations can be made for the short term pressurisation process which occurs within the containment system if a large rupture of the coolant system should occur. Here again, the largest possible breaks dominate the determination of local structural loads and the reaction forces of the primary system components. Again, insufficient knowledge of critical mass flow either costs money in excessive conservatism or yields an insufficient analytical design basis for the structural mechanics considerations.

Another problem which requires an accurate knowledge of the two-phase flow under choking conditions is the behaviour of closing valves under two-phase flow conditions. Several failures of valves (safety valves, check valves) have been observed in the chemical and nuclear industry. Such failures may occur due to the effect of water hammer in a strongly decellerated low quality two-phase flow. In recent years several research activities have been devoted to studying this problem in relation to reactor safety. Carefully designed damping mechanisms for check valves help to avoid such kind of failures.

3

SURVEY OF AVAILABLE CRITICAL FLOW MODELS

This chapter provides a survey of the critical flow models currently available. As a basis for discussion the emphasis has been placed on the models in use in the RELAP4 codes since they are in general use within the OECD area either directly or in locally adapted form. This chapter consists of:

- discussion of phenomenological aspects of critical flow;
- examination of problems encountered in analysing two-phase flow;
- classification of the models;
- short description of the models used in RELAP4 code;
- comparison of the models;
- presentation of the main conclusions of various authors and recommendations for future work.

As indicated in the foreword, this report is not intended as a source book on critical flow modelling. In the preparation of this chapter a detailed survey of the currently available models was drawn up, from which the information presented here has been distilled. This survey has been published as a technical addendum to this report and is noted in this chapter as Reference /1/.

The symbols used in this chapter are defined in a list at the end of the chapter.

3.1 MAIN ASPECTS OF CRITICAL FLOW

In adiabatic, frictionless, single-phase, compressible, steady flow through a De Laval nozzle the critical conditions arising in the minimum cross-section are well-understood and can be described analytically. Three equivalent statements are generally adopted to define criticality from a physical point of view: a) upstream of the critical section the fluid state does not depend upon "small" changes of the thermodynamic variables existing downstream; b) at the minimum critical cross-section there is at least one wave whose propagation velocity is zero with respect to the duct wall; and c) (in the critical section) the flow velocity is equal to the local isentropic speed of sound.

Only the first of the above definitions directly characterises two-phase critical flow; the other two definitions require precautions when applied to two-phase flow. In fact, sonic speed is related to one phase and it is generally quite different in liquid and vapour; moreover, any perturbation wave induces variations in flow structure which may affect critical conditions. Neither of the last two points has been completely investigated.

Many works concerning two-phase flow pattern description in the region of low flow-rate (much less than critical) are available in literature. No work, in the knowledge of the authors, describes two-phase critical flow patterns; these indeed may depend upon the thermodynamic transformations and the geometric flow paths followed by the fluid itself.

Two other aspects of two-phase flow phenomenology are important:

- piezometric head at the broken pipe-reservoir vessel connection: very different flow-rates result by changing break position from liquid to vapour zone;

- incondensable gas content in the mixture which affects the depressurisation phenomenon in the outlet pipe.

This brief discussion gives an idea of the complexity of critical two-phase flow. In the next sub-section some difficulties in translating the above aspects into mathematical terms are outlined.

3.2 DIFFICULTIES IN ANALYSING TWO-PHASE FLOW

Many models have been developed describing the fluid depressurisation in the outlet pipe (or orifice) both from an integral energy point of view (for example, MOODY 1965), and/or as a bubble coalescence and growth phenomenon (for example, WINTERS-MERTE 1979).

Also the critical flow rate for steady-state conditions (which often is not a good approximation to reality) is a function at least of five independent quantities ($\alpha, \rho_g, \rho_f, w_g, w_f$) the most important of which in a quantitative sense, void fraction, cannot be obtained from balance equations. Moreover, none of the above-mentioned variables is reliably measured in experiments. Although there are theories assuming a frozen composition of the mixture in the outlet pipe, both aforesaid energetic and bubble methods must deal with phase change (and hence with void fraction variation).

The following aspects are unresolved from the quantitative evaluation point of view:

- multi-dimensional effects (especially near geometric discontinuities);

- flow pattern characterisation and its dependence on external and internal (with respect to the fluid) momentum and energy sources;

- non-equilibrium always arises to some degree, even in a short flow path or during a short time, whenever a real fluid changes some of its initial state variables* ;

---

* Equilibrium conditions are a particular case of the general link among quantities related to vapour and liquid.

- mass, momentum and energy exchange between the two phases and between each phase and the exterior;

- non-condensable gases released during depressurisation (especially in multi-component flows);

- influence of geometrical and thermodynamic upstream conditions on critical flow: flow paths into the reservoir vessel, rupture position with respect to liquid level, heat source, etc.

None of the models investigated here takes all these aspects into consideration; moreover the last point makes the formulation of one theory valid for all plant situations extremely difficult.

Other difficulties arising in safety analysis of nuclear plants derive from the evaluation of break area shape and from the effect of the two countercurrent jets meeting in the exit section when the rupture is not double-ended.

Finally, in the general field of two-phase flow, the problem of the pseudo-criticality phenomenon must be mentioned. This possibility has been suggested by Bauer et al. /1/. It consists of the flow-rate remaining constant when downstream pressure decreases, caused by vaporisation which increases pressure losses in the pseudo-critical section while maintaining constant flow-rate. (It may also occur at low flow-rate and pressure differences between upstream and downstream reservoirs.)

3.3 CLASSIFICATION OF THE MODELS

In this section the considerations and the model sub-division of /1/ are reported, though, for compactness, a more qualitative analysis is presented here.

3.3.1 CRITERIA

In the last 40 years, several tens of models have been published on the calculation of the critical flow rate of a two-phase mixture.

In some models there is no theoretical support, rather they are semi-empirical formulas, linking critical flow rate to thermodynamic variables generally representing the fluid state in the pressure vessel. These models adopt non-dimensional empirical coefficients in order to fit the experimental data.

Other models, instead, derive from the solution of a set of two or more (up to 6) balance equations, describing the conservation of mass, momentum, and energy for each phase separately or for homogeneous mixtures.

From a mathematical point of view, it ought to be possible to draw, with suitable simplifying assumptions, from complex theories (e.g. based upon all of the six balance equations) models with a smaller number of equations; this is hardly ever achieved as in most models the result is obtained by introducing evolution laws which are more or less particular or arbitrary.

Moreover, many theories become acceptable for engineering calculations if they are related to a given experimental apparatus and/or to conditions assigned beforehand. On the contrary, if boundary conditions (which are usually simplifying assumptions) are varied, the results sometimes diverge or become unusable.

From this brief outline one can foresee quite different formations of the models and difficulty in group classification.

The following sub-division in four groups has usually been adopted (/2/, /3/):

- Theories which assume thermodynamic equilibrium throughout the expansion
  - homogeneous theories ($K = 1$)
  - non-homogeneous theories ($K \neq 1$)
- Non-equilibrium theories
  - "frozen" theories ($K \neq 1$) but having given value*
  - non-homogeneous theories ($K \neq 1$)

This sub-division is observed in the remainder of this chapter. Nevertheless, other characteristics allowing a deeper comparison and giving an idea of the model applicability are referred to; these characteristics are listed below:

- Model Formulation

- number of conservation equations, i.e. equations of mass, momentum and energy conservation of the flowing system;

- number of state and/or transformation equations: the state equations are those describing the system state through some variables (e.g. p, T, h, s, etc.); the transformation equations define the state change according to given criteria**;

- number of constitutive equations: they are usually empirical equations, derived either from non-dimensional analysis or from assumptions regarding the system behaviour***;

- number of analytical conditions added to the model, sometimes without any physical meaning;

- need for semi-empirical parameters for problem solution.

---

* By "frozen" we mean that the composition at the inlet of the flowing pipe is the same as at the outlet.

** The two types of equations have been combined in order to avoid complicated definitions to distinguish them. Consider, e.g. the equation $ds = 0$, it is at the same time a state equation and it defines an isentropic transformation.

*** For example, $\tau = f(h_f, h_g, p, v)$; $q = f(h_f, h_g, p, v)$ are intended as constitutive relationships.

- **Assumptions (phenomenological aspects and parameters considered)**

- transient phenomena;
- multi-dimensional effects;
- non-homogeneity in pressure vessel;
- heat exchange with exterior;
- pipe length (essentially for friction evaluation);
- orifice (I), nozzle (II), constant pipe area (III), any kind of pipe (IV).

- **Output**

- diagrams or relations linking p_o, h_o and/or x_o to Γ;
- diagrams or relations linking p_e and/or x_e to Γ;
- diagrams giving exit thermodynamic variables as opposed to those in reservoir;
- model applicability*.

3.3.2 LIST OF MODELS BY CATEGORY

The models surveyed for this report have been divided into the four main groups mentioned above. Table 3-1 presents these categories and shows for each model:

- the model's main characteristics such as principal assumptions, the analytical solution method adopted, the type of approach to the problem and the structure of the model; and
- observations on the aim of the theory or on particular results obtained.

This table has been synthesised from tables 3-I and 3-II contained in the technical addendum to this report (Reference /1/) and the model references shown in Table 3-1 refer to the technical addendum.

---

* For the definition of model applicability see paragraph 3.4.1.

TABLE 3-1

MODEL SUBDIVISION		NAME OF THE FIRST AUTHOR AND PUBLICATION DATE	MAIN MODEL CHARACTERISTIC	OBSERVATIONS	MODEL REFERENCES	
					BIBLIOGRAFIC REFERENCE /1/	PARAGRAPH OF /1/ IN WHICH THE MODEL IS DESCRIBED
PERFECT FLUID		Perfect gas	Isentropic flow	This is the classical theory - Flow through De Laval nozzle -	/66/	3.3.2.1
PERFECT FLUID		Perfect gas	Diabatic, no friction	- Flow through cylindrical duct-	/65/	3.3.2.2
PERFECT FLUID		Perfect gas	Adiabatic, friction	- Flow through cylindrical duct-	/65/	3.3.2.2
PERFECT FLUID		Incompressible liquid	Isentropic flow	Motion is possible only by an external (with respect to the fluid) energy source	/65/	3.3.2.3
EQUILIBRIUM THEORIES	HOMOGENEOUS THEORIES	HEM	Homogeneous equilibrium model - Isentropic flow	This is the most simple two-phase flow theory - Present in RELAP-	/9/	3.3.3.1
EQUILIBRIUM THEORIES	HOMOGENEOUS THEORIES	Lahey et al. 1977	"Fluid-dynamic" approach	/	/67/	3.3.3.2
EQUILIBRIUM THEORIES	HOMOGENEOUS THEORIES	Babitskiy 1973	Equilibrium scheme	Substitution of momentum equation with isentropic condition	/68/	3.3.3.3
EQUILIBRIUM THEORIES	NON-HOMOGENEOUS THEORIES	Moody 1965	Slip equilibrium model (energy model)	This is one of the most widely used theories - Present in RELAP	/13/	3.3.3.5
EQUILIBRIUM THEORIES	NON-HOMOGENEOUS THEORIES	Moody 1966	Analysis of the whole depressurization phenomenon	/	/69/	3.3.3.6
EQUILIBRIUM THEORIES	NON-HOMOGENEOUS THEORIES	Fauske 1964	Slip equilibrium model (momentum model)	/	/30/	3.3.3.7
EQUILIBRIUM THEORIES	NON-HOMOGENEOUS THEORIES	Levy 1965	Lumped model	The functional link between 'x' and 'α' is obtained supposing the same pressure losses for liquid and vapour phases	/12/	3.3.3.8
EQUILIBRIUM THEORIES	NON-HOMOGENEOUS THEORIES	Cruver et al. 1967	Unified theory of one dimensional isentropic equilibrium separated two phase flow	The authors define and calculate two phase specific volume averaged with respect to different physical quantities	/70/	3.3.3.9

TABLE 3-1 CONTINUED

EQUILIBRIUM THEORIES	NON-HOMOGENEOUS THEORIES	Ogasawara 1969	Eigenvalues method approach to critical two phase flow	/	/72/	3.3.3.10
		Ogasawara 1969	As above	Flow through orifices	/74/	3.3.3.11
		Malnes 1977	This author shows the differences in two phase mixture maximum velocity when the assumptions of frozen flow and equilibrium flow are separately applied	This is not a two phase critical flow model	/85/	3.3.3.12
		Adachi 1973	Two independent energy equations method	The formulation of this theory is different from those shown so far	/4/	3.3.3.13
		Adachi	As above - Flow from cylindrical duct	As above	/6/	3.3.3.14
		Adachi 1974	As above - Flow from orifice	Study of contraction coefficient (C_D)	/7/	3.3.3.14
		Moody 1975	Consistent slip model	/	/9/	3.3.3.16
		Castiglia et al. 1979	Maximum entropic flow through the whole expansion	Results nearly as Moody 1965	/75/	3.3.3.17
		Tentner et al. 1978	Method of characteristics	The variable time appears in the equations	/10/	3.3.3.18
		Wallis et al. 1978	Isentropic stream tube model	The authors by-pass the difficulties connected with slip evaluation	/76/	3.3.3.19
		Ransom et al. 1978	Method of characteristics	Used in the recently developed code RELAP5/MODO	/78/	3.3.3.20

TABLE 3-1 CONTINUED

NON-EQUILIBRIUM THEORIES	"FROZEN" THEORIES	Burnell 1974	Semi-empirical correlation valid in sub-cooled region	Very easy to use	/8/	3.3.4.2
		Zaloudek 1963	As above	As above	/18/	3.3.4.3
		Starkman et al. 1964	Classical "frozen" theory	Based upon same authors experimental results	/19/ /64/	3.3.4.4
		Moody 1969	Pressure pulse model	The author writes balance equations for a moving control volume in which a pressure pulse travelling counterflow is present	/79/	3.3.4.5
		Henry et al. 1971	The purpose is to evaluate critical flow only by knowledge of stag. cond., accounting for non-equilibrium	Present in RELAP	/20/	3.3.4.6
		D'Arcy 1971	Sonic model solution obtained through "small" perturbation method	/	/80/	3.3.4.7
		Ardron et al. 1976	Analogous to Moody 1965	The authors evaluate what they call "upper bound flow"	/55/	3.3.4.8
		Ransom et al. 1978	Methods of characteristics to solve the problem - six equations	Used in RELAP5/MODO	/78/	3.3.4.9
	GENERAL THEORIES	Henry et al. 1970	Model valid in low quality ($x \lesssim 0.02$) region	Simple theory taking into account metastable effects	/82/	3.3.4.11
		Henry 1970	Model valid for high L/D ratio ($L/D \simeq 12$)	The author gives an ineresting interpretation of two phase flow phenomenology - Distinction between smooth and sharp entrance	/31/	3.3.4.12
		Henry 1970	Model valid for very high L/D ratios ($L/D > 12$)		/31/	3.3.4.12
		Klingeibiel et al. 1971	Entrained separated flow (ESF) model	Based upon authors experimental work	/83/	3.3.4.13

TABLE 3-1 CONTINUED

NON-EQUILIBRIUM THEORIES	GENERAL THEORIES	Klingelbiel et al. 1971	Semi-empirical approach	Flow-rate as Moody 1965, with slip evaluated from experimental data	/83/	3.3.4.14
		Malnes 1975	Release of dissolved gases taken into account	/	/84/	3.3.4.15
		Porter 1975	Moody's theory is taken as reference	Analysis performed over all Mollier diagram, from sub-cooled liquid to superheated vapour	/89/	3.3.4.16
		Rivard et al. 1975	Two field, two-phase model	Analysis of the whole depressurisation phenomenon	/90/	3.3.4.17
		Kroeger 1976	Drift flux approximation model	Analytically consistent analysis	/32/	3.3.4.18
		Bouré et al. 1976	Complete one-dimensional two-phase flow analysis	/	/29/	3.3.4.19
		Avdeev et al. 1977	Original method in determining vapour formation rate	Flow from cylindrical ducts	/92/	3.3.4.20
		Travis et al. 1978	Analysis of different questions inquired at present in two-phase critical flow field	This is not a critical two-phase flow model	/39/	3.3.4.21
		Tentner et al. 1978	Method of Characteristics	In contrast with Moody results	/10/	3.3.4.22
		Moesinger 1978	Drift flux approximation assessment	The whole depressurisation phenomenon is considered	/93/	3.3.4.23
		Winters et al. 1979	Lumped non-equilibrium model based on bubble growth analysis	Two empirical coefficients are used	/50/	3.3.4.24
		Romanacci 1976	Simple analysis of vessel depressurisation	This is not a critical two-phase flow model	/53/	3.3.4.25

TABLE 3-1 CONTINUED

THEORIES BRIEFLY DESCRIBED BY GIOT	Katto et al. 1974	Equilibrium isenthalpic flow model	/	/96/	3.3.5.1
	Giot et al. 1968	Method analogous to those of Fauske and Moody	/	/100/	3.3.5.2
	Meunier et al. 1969	Variable slip models	Comparison with Moody and Fauske results	/103/	3.3.5.3
	Giot et al. 1972			/15/	3.3.5.3
	Flinta et al. 1975	Effect of nucleation on critical flow-rate	/	/104/	3.3.5.4
	Bauer et al. 1976	Non-equilibrium model	Pseudo-criticality results theoretically	/105/	3.3.5.5
	Stadtke 1977	Irreversible thermodynamic analysis	Analysis of interfacial transfer terms on critical flow	/109/	3.3.5.6
	Seynhaeve 1977	Analysis of sub-cooled liquid through orifices	/	/110/	3.3.5.7

3.4 OUTLINE OF THE ANALYSIS

3.4.1 GENERAL

In formulating any theory, a certain physical pattern, which more or less reflects reality, is in mind.

In the present case*, the situation shown in Figure 3-1 is taken as reference. In order to obtain the flow-rate it is necessary to consider at least:

- initial thermodynamic conditions in the vessel: enthalpy (H_o), mass (M_o), and pressure (p_o);
- geometrical data: length (L) and diameter (D) of the broken pipe; position of the connection between broken pipe and pressure vessel, with respect to initial liquid level (dimension 'b' in Figure 3-1).

Moreover, in the most general case, a theory should consider all the aspects mentioned in paragraphs 3.1 and 3.2; with reference to a pressure vessel whose volume is not infinite, it becomes necessary to add:

- initial acceleration of the mixture as a function of break opening time, upstream thermodynamic conditions, rupture size;
- link between blowdown flow-rate and pressure history in the vessel.

From Tables 3.I and 3.II and descriptions reported in Reference /1/, it may be observed whether a model covers such aspects or note. The Reference also outlines:

- basic assumptions;
- governing equations;
- model applicability (a theory is "applicable" if it considers, at least, all the variables given in the above as input, and if it shows, as a result, a correlation of the specific maximum mass flow rate (Γ) as a function of initial enthalpy (h_o), pressure (p_o), and/or temperature (T_o) of the fluid in the vessel.

3.4.2 DISCUSSION OF THE MAIN ASSUMPTIONS

The models have been classified in four groups as previously mentioned. In this section a brief discussion of the assumptions common to each group is presented.

---

* In this work only models studying the flow of a two-phase mixture from a pressure vessel without internals are examined (the complex neutronic effects and heat exchange phenomena in a real vessel are not considered).

Thermodynamic Equilibrium Models

The basic assumption of these models is the existence of thermodynamic equilibrium between liquid and vapour phases in each cross-section of the flow duct. By thermodynamic equilibrium we mean that:

- pressures and temperatures of liquid and vapour phases are equal;

- pressure and temperature are linked by the Mollier diagram saturation curve.

The mixture quality is allowed to vary along the duct. Strictly speaking, this is contrary to the thermodynamical hypothesis, because the condensation and evaporation phenomena depend upon differences of temperature and/or pressure between the phases. Implicitly, quality change is supposed to take place at infinite velocity.

It is possible to divide the equilibrium models in two groups:

- homogeneous models: equal vapour and liquid velocity;

- non-homogeneous models: different velocities for the two phases.

Homogeneous models contain the further assumption of mechanical equilibrium between the two phases, i.e. equal vapour and liquid velocities; the mixture density is generally area-averaged.

Non-homogeneous models contain a new variable: the ratio between vapour and liquid velocities different from unity. Consequently, a further equation at least is necessary for problem solution. Often the calculation of the slip ratio distinguishes one model from another.

Moreover, in most of the analysed theories, stationary and isentropic flow is hypothesised. We recall that with reference to a loss of coolant accident, stationary flow is not a "good" assumption in large LOCAs, while the isentropic one is not "good" assumption in small LOCAs.

Non-equilibrium Models

Non-equilibrium models include equilibrium models as a particular case. Non-equilibrium effects arise:

- from the physical fact that two-phase fluid depressurisation velocity may be greater than thermal exchange velocity between liquid and vapour;

- from energy (heat) and momentum (friction) exchange with the duct walls.

According to several experimental findings (see, for example, in /1/ references 26 and 124) the time during which the mixture remains in a metastable state is about one millisecond. Starting from such a value Moody calculated the negligible disequilibrium for pipe length greater than 12 centimetres.

According to other researchers (see Reference 31 of /1/), instead, the degree of disequilibrium depends on the L/D ratio: i.e. non-equilibrium phenomena must be considered for $L/D \leq 1/4$.

However, even in long channels it probably plays an important role on thermodynamic conditions at the exit section. Besides, all researchers seem to agree that thermal non-equilibrium can appreciably increase both the critical flow-rate and the propagation velocity of rarefaction waves normally generated at the ruptured section.

Many theories treating non-equilibrium have been developed: some of them are no more than empirical formulas for rapid calculation of maximum flow of initially sub-cooled water; others make use of very sophisticated models.

Most of these theories introduce empirically determined coefficients: the latter include the number of vaporisation nuclei per unit mass of flowing mixture, a constant relating certain non-equilibrium quantities to equilibrium values, a coefficient that is needed to take into account a certain assumed flow contraction, experimentally observed in a single-phase flow in orifices.

The feature which is common to all these coefficients is that they are obtained by matching experimental data to theoretical results.

As already indicated (paragraph 3.3), the models are subdivided in two groups:

- frozen theories (which admit no variation in mixture composition while flowing);

- non-homogeneous, non-equilibrium theories (in which no implicit assumption is made on the link between pressure and temperature of of the two phases*).

The majority of non-equilibrium models have been developed in recent years.

Regarding "frozen" theories, the assumptions generally adopted are:

- the velocity ratio between the two phases is given;

- no heat or mass transfer takes place between the phases; thus the quality remains constant throughout the expansion (this assumption characterises "frozen flow");

- the vapour expands according to an assigned law: for example, according to gas dynamic principles.

All vapour and liquid changes are then assumed to be independent from one another, as far as the exit section at least is concerned.

According to the experimental results, a "frozen" theory is an accurate representation of reality when flow transit time in the outlet pipe is less than the time constant of the phase change process. This constant, among other things, is not easy to determine as it depends mainly upon liquid-vapour exchange area (and hence flow pattern), number of nuclei per unit volume and smoothness of duct surface wall.

---

* We define a theory as non-equilibrium when at least one thermodynamic quantity does not follow the saturation line.

It is instructive to see the qualitative and quantitative differences between mixture velocities in a homogeneous equilibrium theory and a frozen theory (see Figure 3-2).

3.5 DESCRIPTION OF MODELS IN RELAP4

In this section, reference is made to the RELAP4/MOD5 user's manual /4/. Four junction critical flow models have been incorporated into this code. They are:

- sonic critical flow model;

- MOODY model (1965);

- HENRY-FAUSKE model (1971);

- homogeneous equilibrium model (HEM).

The procedure automatically followed by the code consists in calculating at each junction the flow-rate with the user-selected critical flow model and the simple inertial flow: the minimum of the two values is adopted as the appropriate junction value.

Here these models are briefly described (in /1/ a more complete analysis is reported).

3.5.1 SONIC MODEL

(See also paragraph 3.3.3.12 of /1/.)

By assuming a homogeneous mixture in thermodynamic equilibrium in the upstream (with respect to the junction) control volume, the code calculates a sonic velocity adopting the classical formula $a^2 = [dP/d\rho]_s$. The input data are the upstream volume conditions: specific internal energy and specific volume are the independent variables. The mass flow is obtained by the continuity equation $[W = \rho_m aA]$. The junction density (ρ_m in the last equation) is firstly assumed equal to the upstream volume density, then it is modified taking into account frictional losses and kinetic energy change up to the junction. The mixture is then expanded isentropically from the upstream volume Mach number to a Mach number equal to one in the junction.

The Model takes into consideration, too, the possibility of a perfect gas dissolved in the water, giving a more complicated expression for critical velocity.

3.5.2 MOODY MODEL

(See also paragraph 3.3.3.5 of /1/.)

This is a one-dimensional, thermodynamic equilibrium two-phase flow theory.

It has been one of the most widely known and accepted models for critical two-phase flow; the USNRC considered it to be the best approach in predicting critical flow rate from long channels (L/D >> 1).

Additional assumptions are:

- uniform axial velocity for each phase in each cross-section;
- isentropic flow: no shear stresses either between the phases or between each phase and duct wall;
- stationary flow.

MOODY has one continuity equation and one energy conservation equation: by combining them and considering definitions of mixture enthalpy and entropy, area-averaged specific volume and slip ratio, the flow-rate is obtained as a function of local pressure and slip ratio (the two variables are assumed to be independent of each other). At this point MOODY assumes that the critical flow rate is the analytical maximum of this function so that $W_c = \max f(p,x)$, a function only of the upstream volume conditions.

3.5.3 HENRY-FAUSKE MODEL

(See also paragraph 3.3.4.6 of /1/.)

The aim of this model is to solve the problem of two-phase critical flow using only a knowledge of the stagnation conditions and, at the same time, accounting for the non-equilibrium nature of the flow.

The duct considered is a De Laval nozzle.

The authors write mass continuity and momentum conservation equations; in the latter, wall shear stresses are assumed to be negligible by comparison with the forces due to momentum variation and pressure gradient. The other assumptions are:

- critical flow rate independent of the pressure at the throat;
- quality constant up to the throat;
- liquid temperature, vapour and liquid entropies at the throat equal to the corresponding reservoir quantities;
- slip ratio equal to unity;
- polytropic expansion of the vapour at the throat;
- at the throat: $[dS/dp]_t = 0$ and $[dx/dp]_t = f[dx_E/dp]_t$ where x_E is the equilibrium quality and f is an empirically determined function of x_{Ec} (f characterises this model).

Combining the above equations and assumptions, critical flow-rate is obtained as a function of throat pressure. A new equation is then necessary; this is obtained by integrating the momentum equation between reservoir and throat conditions.

It should be noted that:

- the whole formulation is related to the low quality region;
- the model assumes neither completely frozen, nor complete equilibrium heat and mass transfer processes;

- the authors recommend a value of 0.84 for the critical flow-rate reduction coefficient in predicting results for orifices or short tubes.

3.5.4 HOMOGENEOUS EQUILIBRIUM MODEL

(See also paragraph 3.3.3.1 of /1/.)

As in the sonic model, all the assumptions characterising homogeneous equilibrium flow are present. HEM differs from the sonic model only in the way it is utilised by RELAP4 so that no other investigation is required here.

Instead it may be interesting to note that:

- this is the simplest theory which may be formulated in studying two-phase flow;
- it was initially developed for the analysis of situation in which the ratio outlet pipe diameter/reservoir vessel diameter is much less than unity.

3.5.5 DISCUSSION

As already said, in RELAP calculations it is possible to use combinations of the above models for analysing the same transient. One of the most used is the combination of HENRY-FAUSKE and HEM models: up to the junction quality value of 0.01 - 0.02, HENRY-FAUSKE is adopted, for greater vapour content HEM is used.

A quantitative analysis of the above models is given in /4/ from which it is seen that:

- the inertial model depressurises a system most rapidly and exhibits the highest sub-cooled flow rate;
- the majority of the flow models and their combinations, with the exceptions of the intertial model, the sonic model and the HEM, exhaust about the same amount of water from a system in the same blowdown time;
- the differences among the models lie in the paths taken in the $\Gamma - x$ plane: some exhibit a higher flow rate during sub-cooled conditions, while for others the reverse is true.

Finally, it may be observed that:

a) the manual gives no indication about the range of validity of each model;

b) apart from the very simplifying assumption of isentropic flow, the interrelation between critical flow rate, friction and mixture evolution is not emphasised;

c) the use of model combinations has no physical meaning: it increases the possibility of matching experimental data (when these are available), it increases the scatter in the computed results, and it decreases the code reliability in predicting new situations.

One goal for future code development should be to limit, as far as possible, the user available options and 'dials' on critical flow models in codes and to give deeper and clearer information concerning the applicability of the model to real situation.

3.6 COMPARISON OF MODELS

3.6.1 GENERAL

Before drawing conclusions it should be pointed out that not all models presented in the bibliography of /1/ have been taken into consideration*. However, some general conclusions can be obtained from the models described above. In particular, it may be observed:

1) probably in no other engineering field do we have such a variety of number and type of equations used to solve the same problem (namely the critical two-phase flow-rate);

2) the importance attributed to the physical aspects, mentioned in part B of the Table 3-1 /1/, is different from author to author;

3) very few authors give results which can be easily compared with others; i.e. there is not a generally accepted format to show the results. Such a format could be, for example, a diagram showing Γ_e versus p_o with h_o as parameter or Γ_e versus p_e with x_e as parameter;

4) the philosophies of approach to the problem and the methods of mathematical solution are in number more or less equal to the number of models;

5) most models presented do not have an explicit expression for the flow rate. From the few cases in which the analytical expression for Γ_e is given, the difference in results between the models easily follows from Table 3-2 /1/. Such a fact also results from many figures reported in the same work;

6) the uniqueness in development of each physical model leads to incompatibilities with the various existing two-phase theories: i.e. it is not possible, starting from a single general theory and making appropriate assumptions, to arrive at all existing theories.

From the above points, it is easy to see the difficulties in comparing the models. Moreover, we think it is useless to show for instance that a model predicts a flow rate greater than another at low reservoir quality and an opposite behaviour at high quality, since each model is generally considered valid only in a particular zone of the Mollier diagram and since such a comparison does not reach any concrete conclusion with regard to the absolute validity of the model itself. Notwithstanding this statement, we will show some differences between the models in the next section.

---

* With regard to this, it is estimated that about 20% of models are missing from this analysis.

A better procedural method for the comparison of the models could be to take as a reference one or more experimental results in which all the measured quantities (especially flow rate) are clearly given with their respective uncertainties, and then to apply the models in questions. Unfortunately, it has not been possible to perform such work for lack of time.

3.6.2 QUANTITATIVE COMPARISON

Many researchers have performed a comparison of two or more models (see, for example, references 27, 28, 55, 115 in /1/). Some results, given in reference /1/, are reproduced here in Figures 3-3 through 3-11, the results obtained using different models may be identified through the bibliographic references of Table 3-1.

The main conclusions drawn are as follows:

- the quantitative differences between the results obtained by some models (see Figures 3-3, 3-4, 3-7) are unacceptable from an engineering point of view;

- the worst discrepancies are related to the low quality region and to low values of L/D ratios ($L/D \leq 8$);

- sometimes models give very different results when a parameter which is not clearly chosen by the authors is varied; the slip ratio is one of such quantities and Figures 3-6 and 3-8 show its influence on critical conditions.

Many other observations analogous to the above may be obtained by analysing all the figures presented in the reference but this is useless without the support of experimental data. Moreover, it is difficult to isolate the causes or the quantities leading to such discrepancies.

When studying the most fully developed models (paragraph 3.3.4.17 and subsequent /1/) the difficulties of a comparative analysis are even greater both for the lack of standard results in terms of flow-rate, and for the great number of variables involved in calculations.

It may be interesting to note the type of comparison adopted by Wallis /5/ and by Hall /6/ who studied the behaviour of different models over a whole blowdown transient. In particular, Hall studied the output of calculations performed with five different models both in terms of flow rate and of exit pressure, and he concluded that none of them was satisfactory from both points of view. In Figures 3-10 and 3-11 we report the results obtained by Wallis compared with experimental values measured by Hutcherson /1/.

Finally, in Figures 3-12 through 3-15, we present some results obtained at the Nuclear Institute of Pisa University in applying RELAP4 to the Standard Problem 4 /7/ (test S-02-6). In these figures the unbroken lines represent the experimental values of exit flow-rate, exit density, upper and lower pressures, respectively, all versus time; the dotted lines show the results obtained from RELAP4 MOD2 adopting the MOODY critical model, and the dashed lines show the result obtained from RELAP4 MOD5 adopting the combination of the HENRY-FAUSKE/HEM model. One can observe the enormous discrepancies especially in flow rate prediction, while the pressure trends are nearly independent of the choice of critical model. It is also to be noted that, in the context of this work, the curves are valid only over the first few tenths of seconds; in fact, after about 50 seconds the trends depend mainly upon volume number and disposition. Analogous results, not shown here, are obtainable by changing the value of the vena contracta coefficient C_D.

3.7 CONCLUSIONS ON AVAILABLE MODELS

The use of a model depends upon the aims and resources of the user and upon the predicted range of variation of some important variables. For example, in a first approximation analysis, the use of a simple theory (as MOODY, HENRY, HENRY-FAUSKE, HEM, etc.) may be suitable, while in order to obtain more reliable results a more sophisticated analysis must be used, always taking the author's assumptions well into account. In the two next sub-sections we will show the main conclusions of some authors and give a brief discussion of the results of this chapter.

3.7.1 MAIN CONCLUSIONS OF SOME AUTHORS

OGASAWARA 1969 /1/

- "Eigenvalue method defines criticality as a discontinuous condition in the flow differential equations systems, and can treat easily a complicated system composed of, for example, separate momentum between two phases, energy and mass conservation equation".

- "Critical condition with energy conservation gives a more accurate solution than the one with entropy conservation".

MOODY 1975 /1/

With reference to theoretical predictions of the two-phase equilibrium discharge rate from nozzles on a pressure vessel he says that:

- "Theory which predicts critical flow data in terms of pipe exit pressure and quality severely over-predicts flow-rate in terms of vessel fluid properties. This study shows that the discrepancy is explained by the flow pattern".

In his conclusion, MOODY discusses many phenomenological aspects influencing flow rate.

MALNES 1975 /1/

- "Critical conditions in a one-dimensional sense are not obtained in two-phase flow, due to two-dimensional effects".

- "Critical mass flow may, however, be calculated from the one-dimensional continuity equation utilising a flashing correlation as mass flows are not sensitive to outlet conditions".

- "Gas content seems to be an important parameter in critical two-phase flow".

ARDRON-FURNESS 1976 /1/

After comparing different models (see Figures 3-3, 3-4, 3-7) with experimental data these authors conclude as follows:

- "There is an absence of a general critical flow theory which satisfactorily describes observed effects of outlet geometry and discharge flow rates, and can thus be confidently applied to reactor blowdown calculations".

KROEGER 1976 /1/

- "Treatment of the phase change front as a discontinuity similar to the treatment of shocks in single-phase gas dynamics permits very accurate solutions".

BOURE-FRITTE-GIOT-REOCREUX 1976 /1/

- "In studying two-phase flow, since the evolution of the mixture is, in fact, a consequence of the transfers at the wall and at the interface, it is more rational to postulate transfer laws than to assume mixture evolution".

- "The mathematical form of the above transfer laws is of a primary importance and it is proposed to allow for the presence in the transfer terms of partial derivatives of dependent variables

TENTNER-WEISMANN 1978 /1/

- "A simple fluid model incorporating slip and coupled to the method of characteristics appears to provide a useful technique for analysis of two-phase flow behaviour".

- "Choked flow in long pipe lines can be predicted from an homogeneous equilibrium model" (many other authors agree with this observation); "discrepancies are observed when this model is used to compute pressure at pipe exit".

- "Thermodynamic non-equilibrium at the exit, rather than high slip ratios, may cause inconsistencies arising in the vicinity of a break".

- Finally they plead for carefully controlled experiments in order to check proposed models.

MOESINGER 1978 /1/

- "The essential blowdown quantities, namely flow rate and pressure history, are described well enough by the drift flux approximation".

- "Phenomena which are only poorly described by the drift flux model, like phase separation, and the behaviour of the separated phases during the rapid transient stage of the blowdown, play an inferior role because they do not influence the behaviour of the overall blowdown process".

- "For other problems where phase separation and the flow fields of the single phases are really of interest, a six equation model which takes the inertia of the droplets or bubbles into consideration should be used".

TRAVIS-RIVARD-TORREY 1979 /1/

- "The success in comparison of experimental and theoretical data has instilled confidence in the predictive ability of the new vapour production model".

WINTERS-MERTE 1979 /1/

- "The non-equilibrium model predictions were considerably more accurate than those of the phase equilibrium model, although consistent discrepancies were observed in all model data comparisons".

- "It is believed that two-dimensional effects near the pipe exit, liquid inertia influences on bubble growth, and limitations in the choked flow model are primarily responsible for the discrepancies".

RANSOM-TRAPP 1980 /1/

- "Comparison of the RELAP5/MODO calculations with the Marviken III Test 4 provided a good evaluation of the ability of a two-phase thermal-hydraulic model to predict correctly mass discharge rates under choked flow conditions at large scale; this both in the sub-cooled and low quality flow regime".

WALLIS 1979 /5/

In this recently published work Wallis compares different models with experimental data. His main conclusions are:

- "The HEM is not a bad way of predicting critical flow in long pipes where there is sufficient time for equilibrium to be achieved and when the flow pattern is conducive to interphase forces that are adequate to repress relative motion. Errors can be large for short pipes (a factor of 5) and significant in longer pipes (factor of 2) if the flow regime, such as annular flow, allows large differences in phase velocities".

After having analysed models considering nucleation processes, Wallis concludes:

- "Although various authors make various assumptions about the source of bubbles, none is able to escape entirely from empirism: it is either blatant or hidden somewhere in the algebraic derivations; the numbers are chosen to correlate the resulting critical flow behaviour rather than to represent measured nucleation characteristics. Thus, the increase in realism gained by incorporating this process into the analysis is largely offset by uncertainties about the quantitative mechanism involved".

He arrived at a similar conclusion by analysing vapour generation models. With regard to "two fluid" models (six equation models) he says that this is undoubtedly the best approach to the problem even if some questions (e.g. the transfer laws) remain. Moreover, he points out that attempts to reduce the full two fluid model to a smaller number of equations are not generally successful*.

---

* On this subject he says that drift flux approximation approach is ill-advised.

Finally Wallis concludes:

- "I believe it is a good rule that the sophistication of a theoretical analysis should match the degree to which the physical phenomena can be specified".

BOURE 1978 /1/

After having reviewed critical phenomena with particular reference to quantitative evaluation difficulties, Bouré concludes:

- "It may be concluded that criticality is fairly well understood, qualitatively. A flow is critical when the displacement velocity in the critical cross-section of the most rapid countercurrent "efficient" disturbance is zero. A flow is "pseudo-critical" when all efficient countercurrent disturbances are damped out in reaching the pseudo-critical cross-section. To enable quantitative prediction of two-phase critical flows to be made, further work is necessary on:

  - nucleation;
  - constitutive laws, including the topological law which controls the pressure difference between the two phases, and the transfer laws, especially in the presence of large gradients;
  - wave phenomena (displacement velocities, damping, dispersion);
  - two- and three-dimensional effects;
  - pseudo-criticality conditions".

3.7.2 RESULTS AND DISCUSSION

Three aspects immediately result from the review of about sixty different models and many other works concerning two-phase flow:

- the great difference in formulation and results we have already spoken about;
- the fact that nearly all the authors obtain good agreement between their theoretical results and selected experimental data;
- the fact that in the development of new models existing work is rarely taken into consideration, so that up to now a uniform research line is not evident; rather there are different and divergent approaches to the problem.

With regard to the differences between the formulations and the results obtained, the main cause is due to the complexity of the phenomena studied and the lack of reliable experimental results. With particular reference to this aspect, it can be noted that almost all researchers feel the need for a deeper experimental understanding of the variables involved in critical two-phase flow.

In the last four years many relatively advanced models have been published (see, for example, bibliographic references 10, 32, 29, 90, 93 and 131 of /1/) but their formulation is far from being unique. A problem common to most of them concerns the analytical model used to describe vaporisation rate.

At this point it may be interesting to distinguish two lines of approach to critical two-phase flow rate evaluation:

a) an engineering-computational approach;

b) a physical-mathematical approach.

In case a) the aims of each model should be the following: to fit the experimental data as well as possible (by adopting empirical coefficients and analytical conditions) and to retain maximum analytical simplicity.

For example the models of HENRY-FAUSKE (paragraph 3.3.4.6 of /1/) and Burnell (paragraph 3.3.4.2 of /1/) are part of this category by adopting empirical coefficients, N and C respectively (in particular these authors follow an engineering approach); the MOODY model too (paragraph 3.4.2) is part of this category by achieving maximum flow rate through analytical conditions with no physical meaning (in particular this author follows a computational approach).

In case b) instead it is necessary to analyse strictly the phenomena and to develop equation systems which enable the engineer to evaluate the main aspects of two-phase flow dynamics without any further reference to experimental results. The criticality condition must be implicit in these models.

This means that the use of empirical coefficients must be excluded, and each analytic expression must correspond to a real situation or at least to a rational assumption about the fluid transformation. In the bulk of the models examined, the two approaches are considered simultaneously.

Moreover, from a fluid-dynamic point of view, a synthesis of the whole LOCA in a LWR, taking into account the effects due to the vessel's internal geometry on critical two-phase flow rate cannot, at present, be handled from a mathematical point of view. Therefore, approach a) seems useful.

On the other hand, simple situations like the one shown in Figure 3-1 should be the target of a model adopting approach b). In this case we think that for a complete knowledge of the whole blowdown phenomenon it is necessary to calculate both the instantaneous value of the pressure in the vessel and the flow rate. Such a model must consider (apart from the aspects we have spoken about in the introduction):

- dynamics of the rarefaction wave generated at the break;

- acceleration of the two-phase mixture;

- flashing phenomena in the pressure vessel subsequent to the arrival of the depressurisation wave.

LIST OF SYMBOLS USED

a	sound velocity
A	flow area
b	hydraulic head
C_D	area reduction coefficient or vena contracta
D	pipe diameter
h	specific enthalpy
H	total enthalpy
K	slip ratio
L	pipe length
M	mass
p	pressure
q	heat exhanged by two-phase mixture
a	specific entropy
t	time
T	temperature
v	specific volume
w	phase velocity
W	flow rate = ΓA
x	quality
α	void fraction
Γ	specific flow rate
ρ	density
τ	wall shear

SUBSCRIPTS INDEX

c	critical value
e	exit
f	liquid phase
g	vapour phase
m	mixture
o	initial value of reservoir value
s	constant entropy
t	throat

Figure 3-1 · **REFERENCE SCHEME FOR THIS WORK**

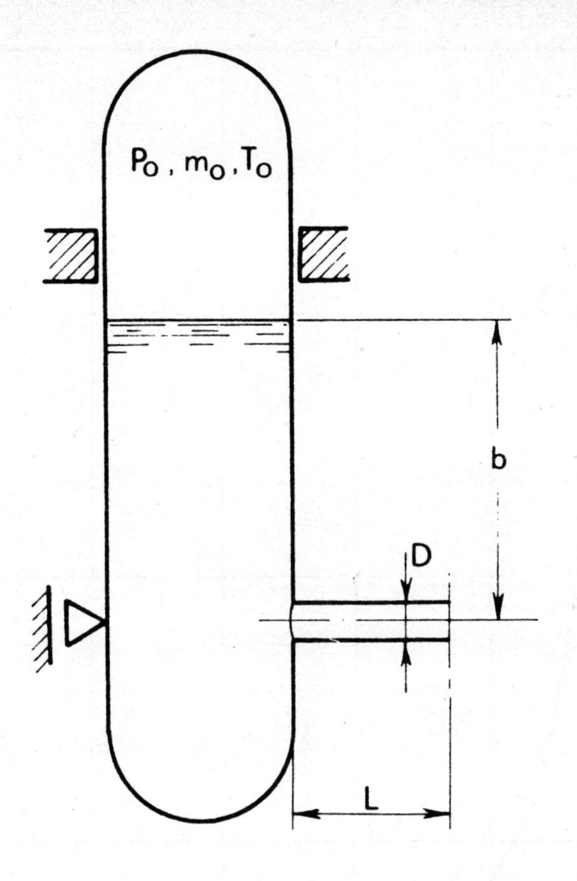

Figure 3-2. **DIFFERENCES IN CRITICAL VELOCITIES FOR TWO APPROACH**

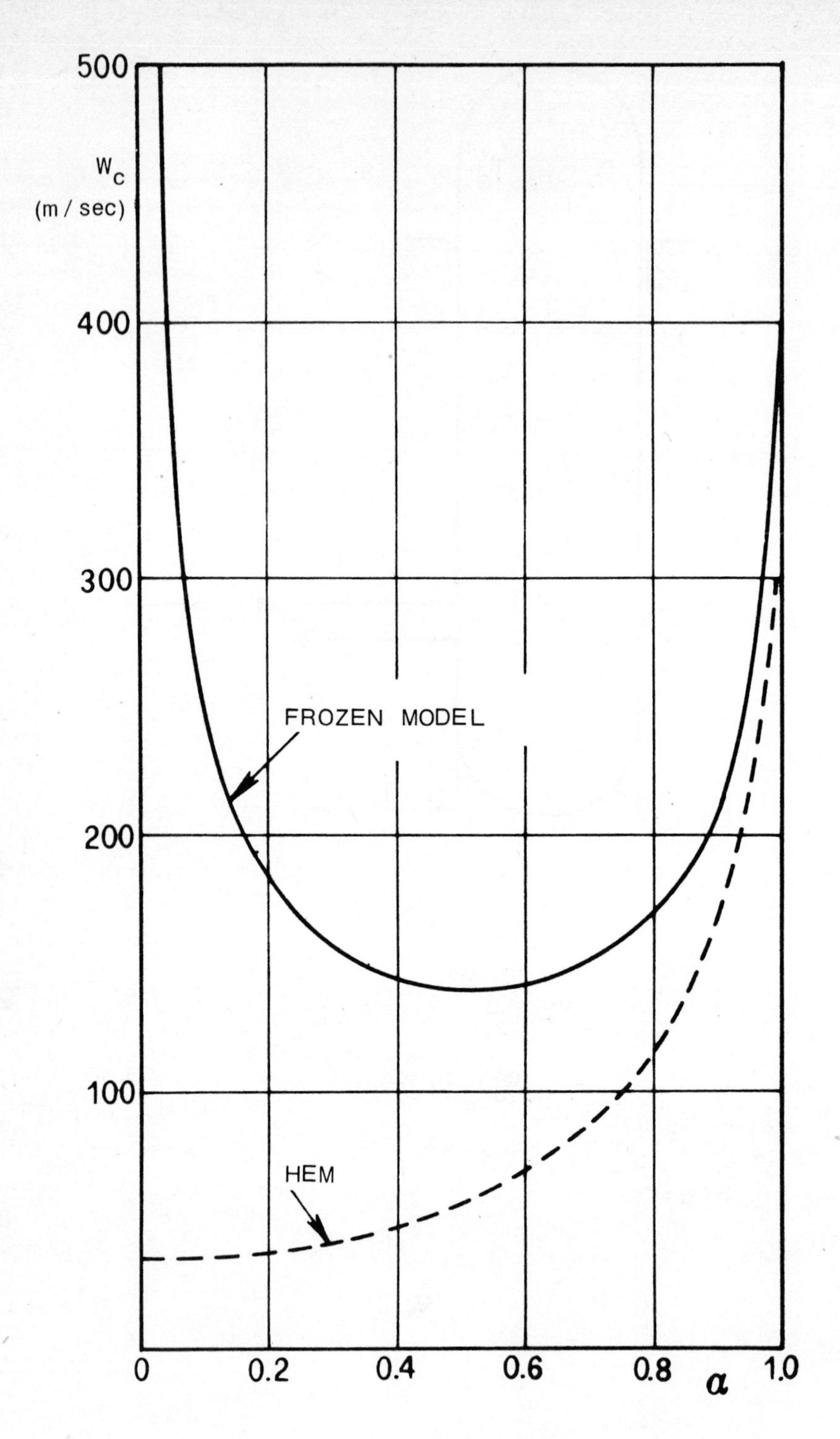

Figure 3-3 **ARDRON ET AL. 1976**[1]

[p_o = 62 bars]

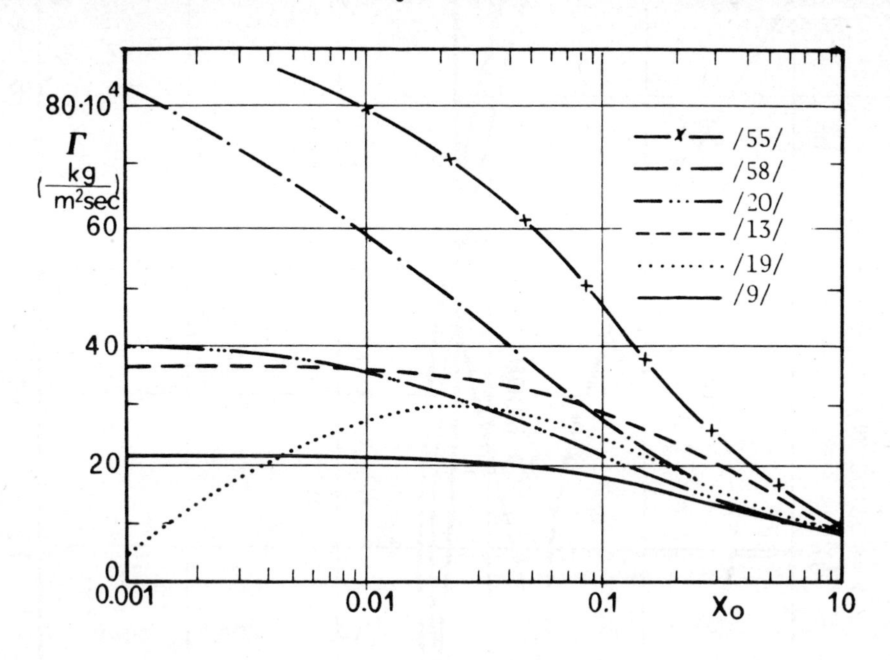

Figure 3-4 **ARDRON ET AL. 1976**[1]

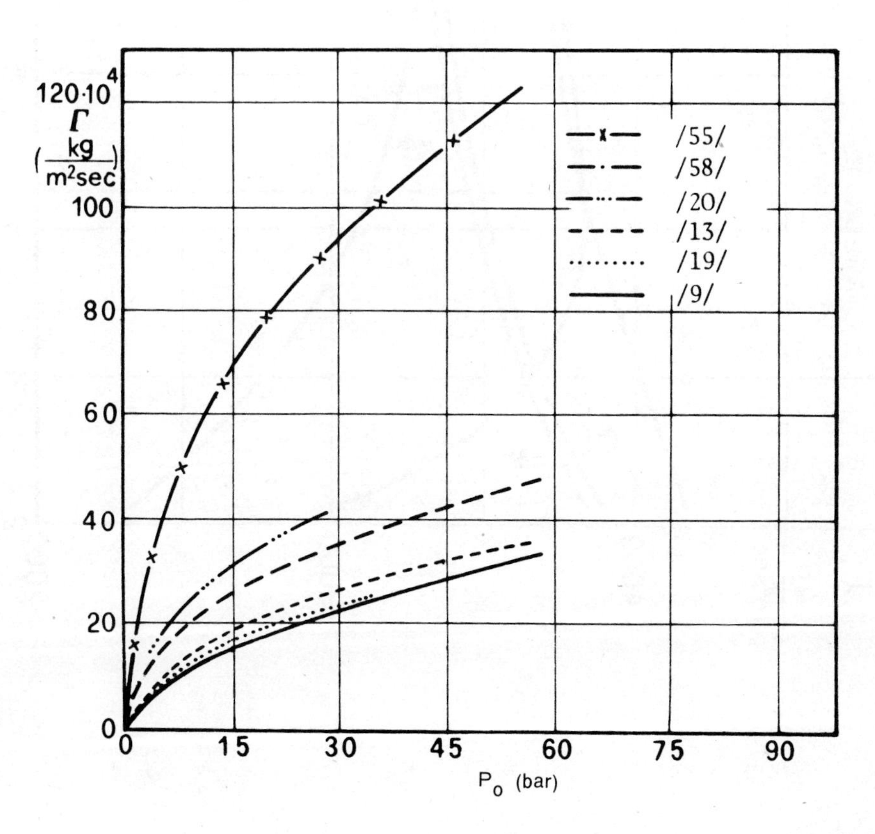

Figure 3-5 **GIOT ET AL. 1972 /2/**

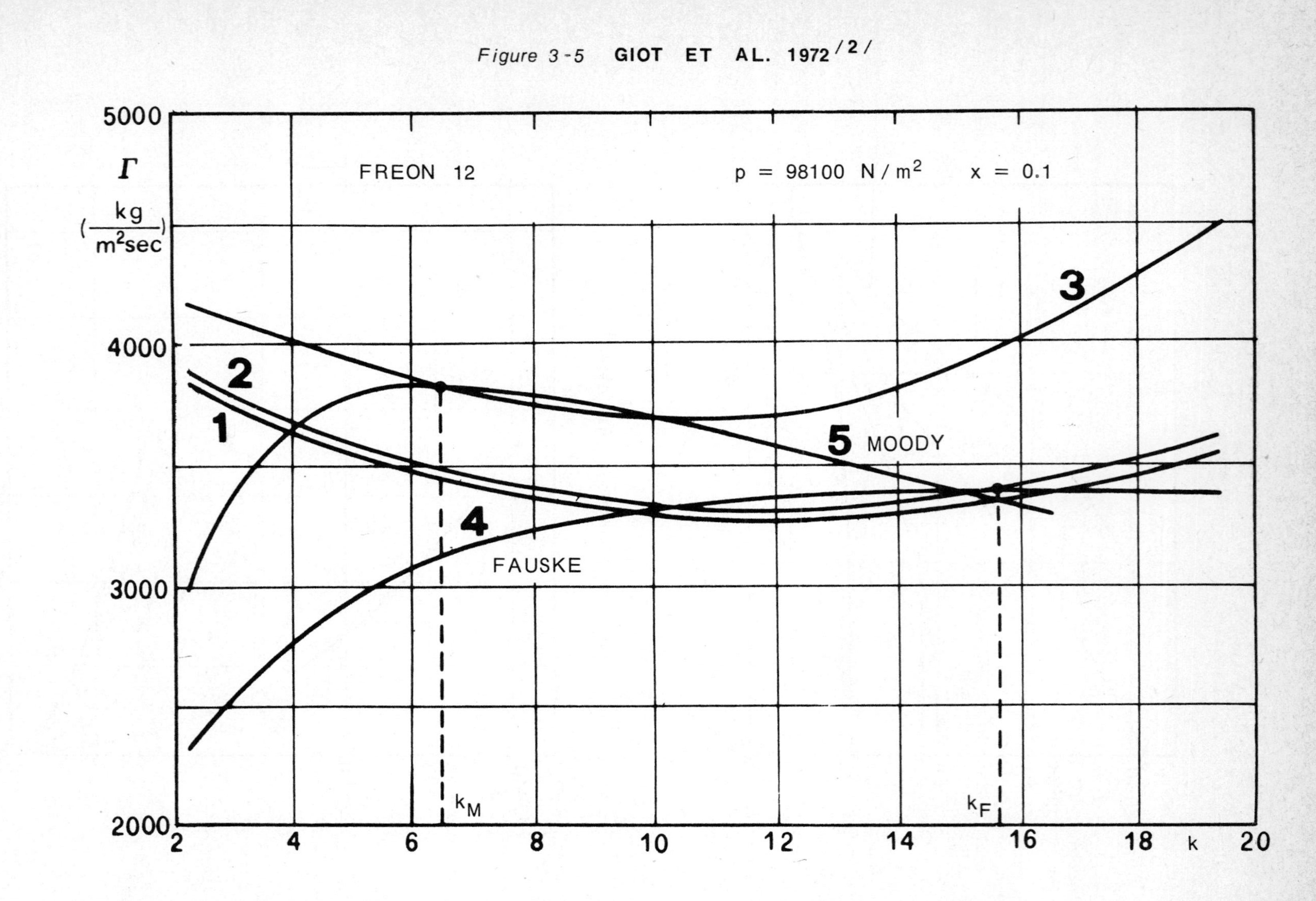

Figure 3-6 **ARDRON ET AL. 1976**[1]

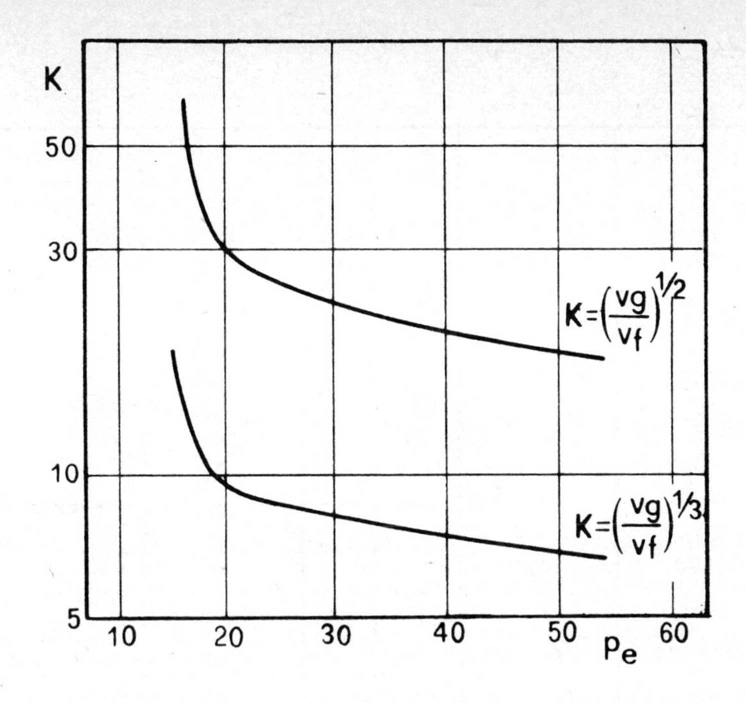

Figure 3-7 **ARDRON ET AL. 1976**[1]

[p_o = 6.2 bars]

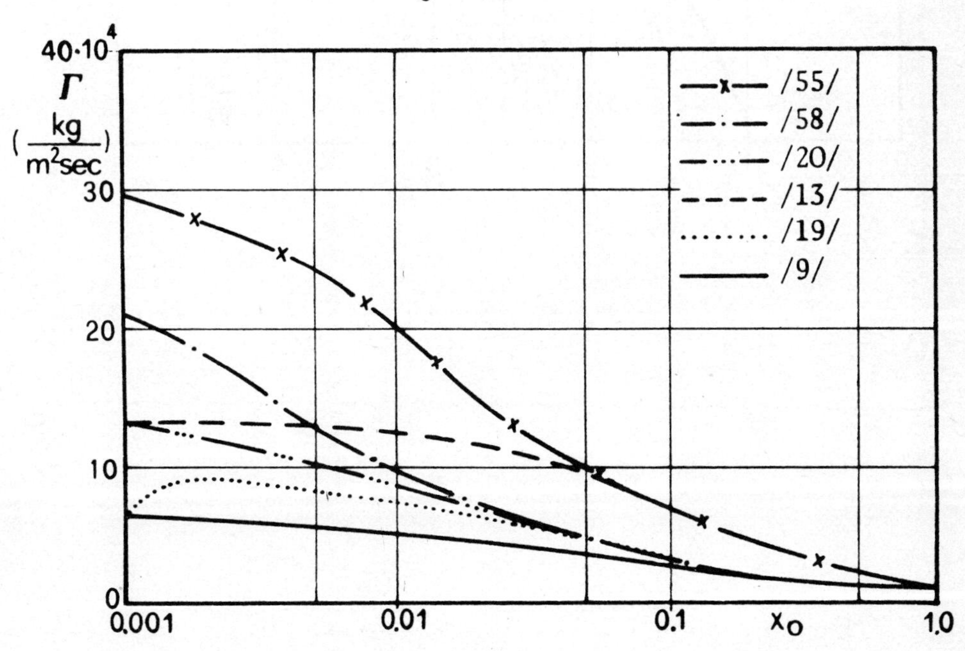

Figure 3-8 **WALLIS 1979** /5/

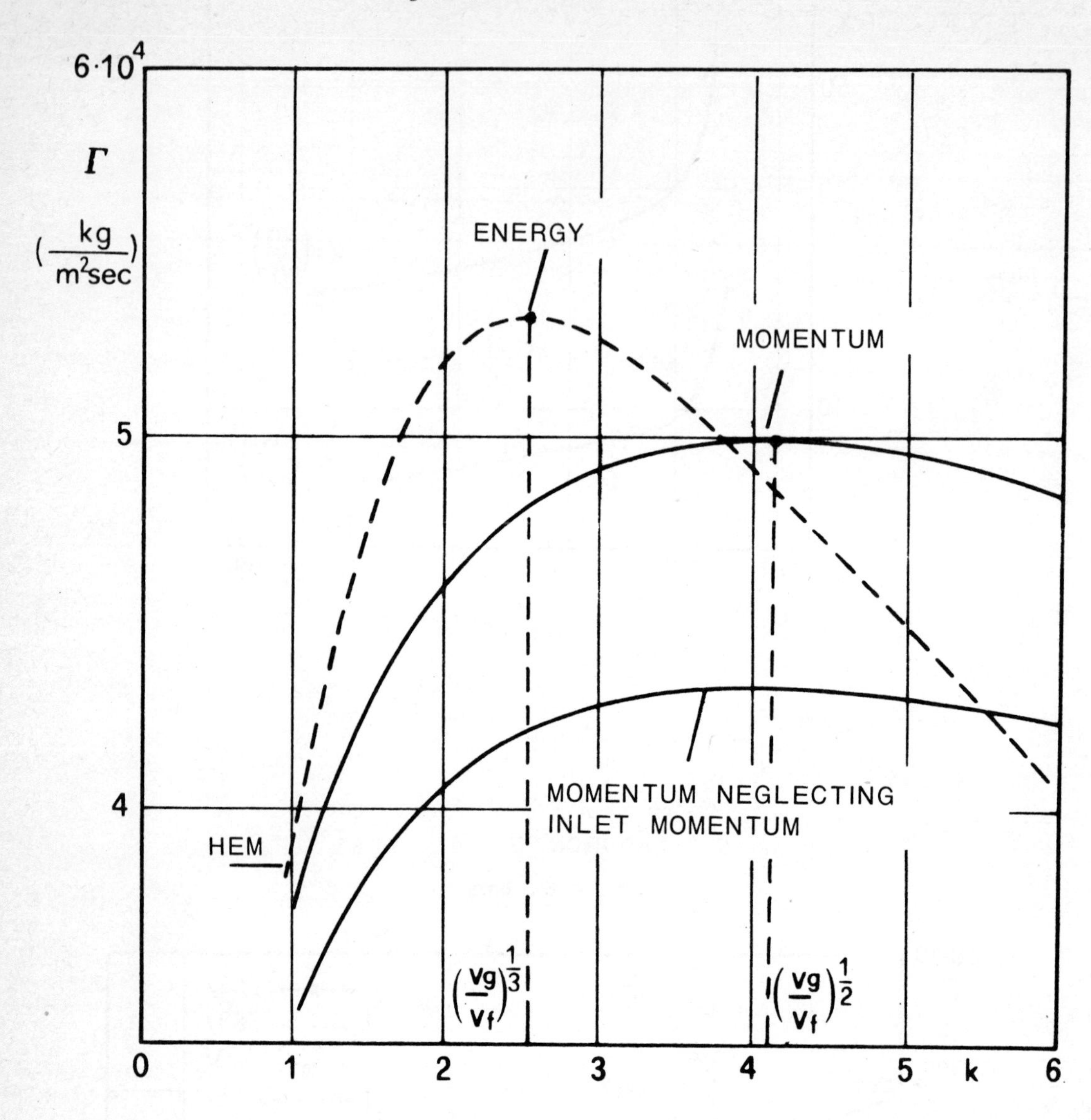

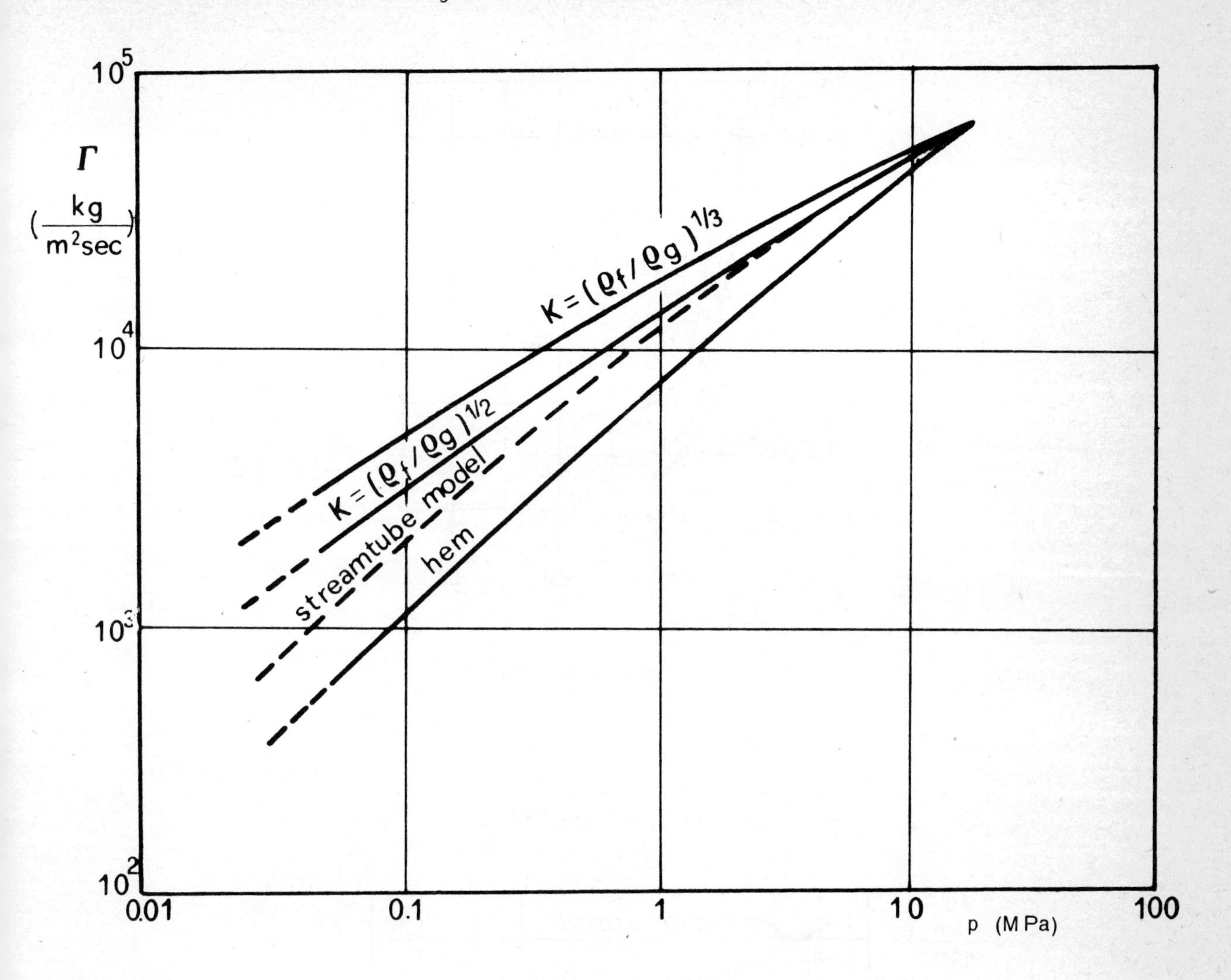
Figure 3-9 WALLIS 1979 /5/
Γ
(kg/m²sec)
10⁵
10⁴
10³
10²
K = (ϱf / ϱg)^1/3
K = (ϱf / ϱg)^1/2
streamtube model
hem
0.01
0.1
1
10
100
p (MPa)

Figure 3-10 **WALLIS 1979**/5/

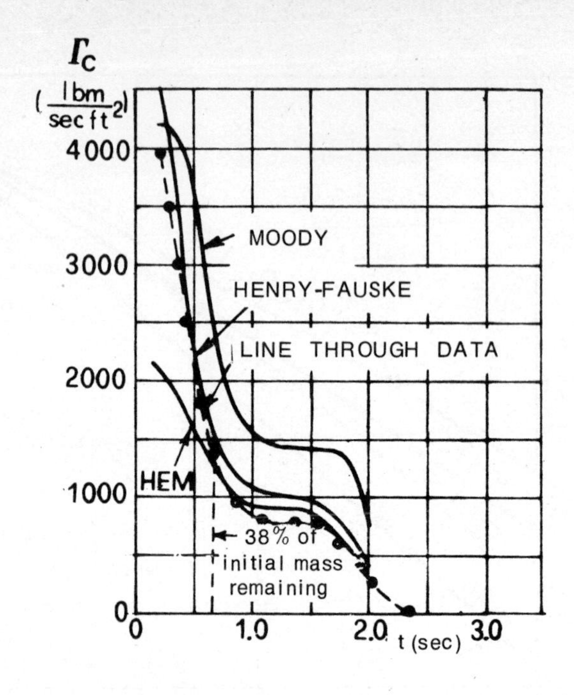

Figure 3-11 **WALLIS 1979**/5/

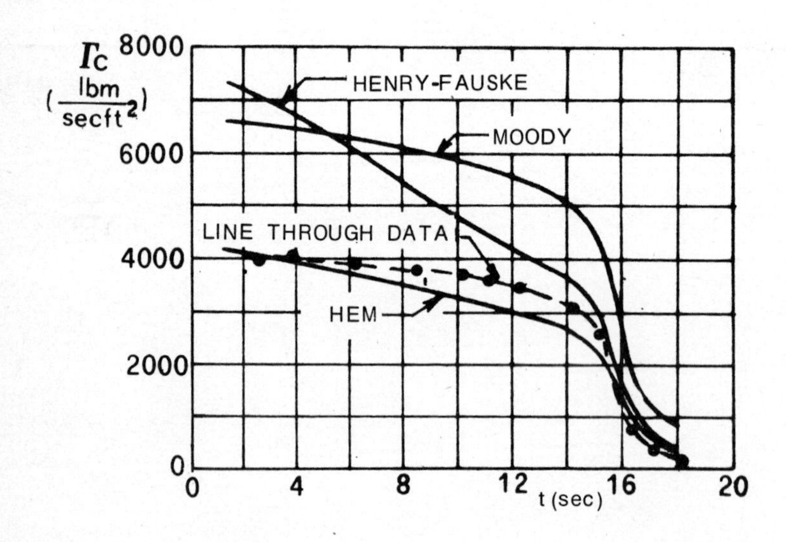

Figure 3-12 **CERULLO ET AL. 1980/7/**
FLOWRATE AT BREAK JUNCTION

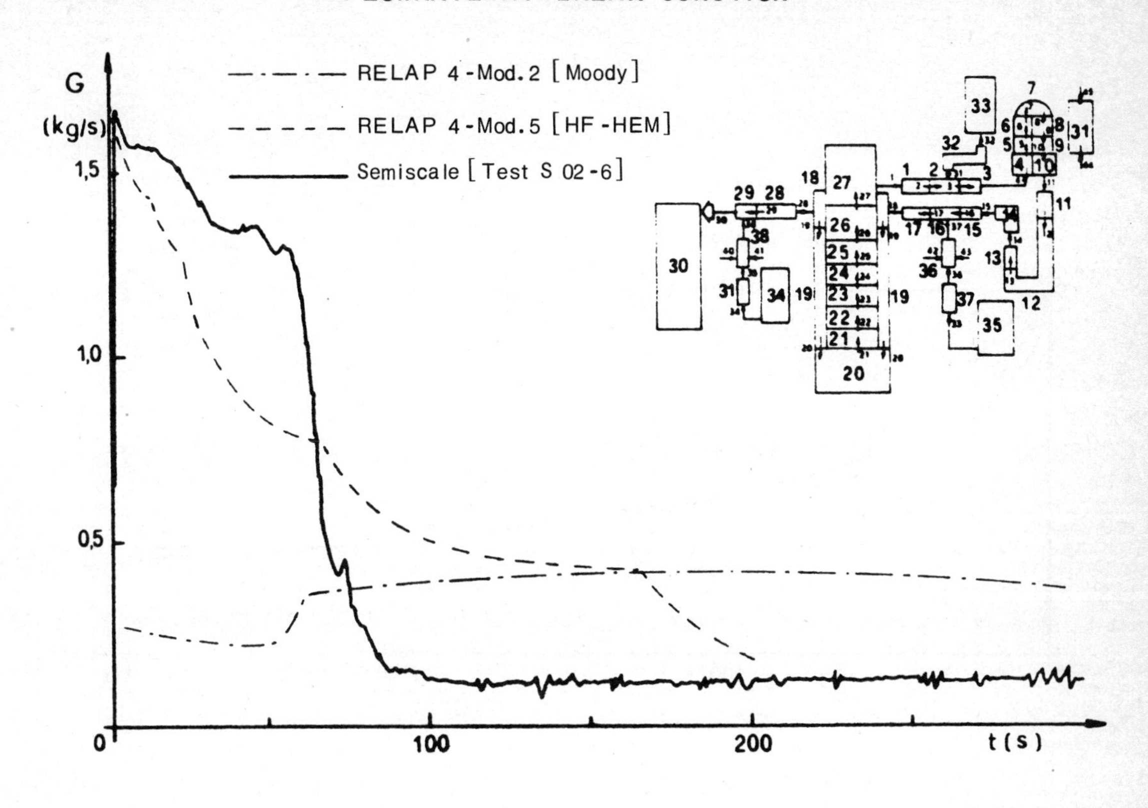

Figure 3-13 **CERULLO ET AL. 1980/7/**
MEAN DENSITY IN VOLUME No. 29

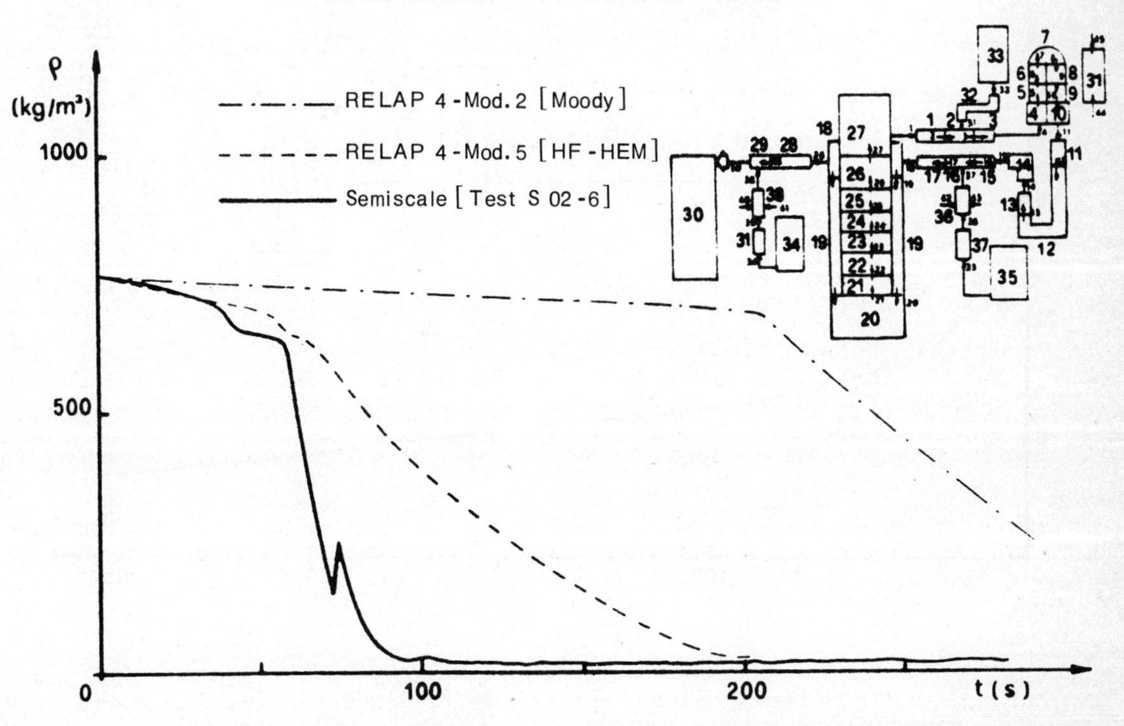

Figure 3-14 **CERUOOL ET AL. 1980** /7/
UPPER PLENUM DEPRESSURIZATION

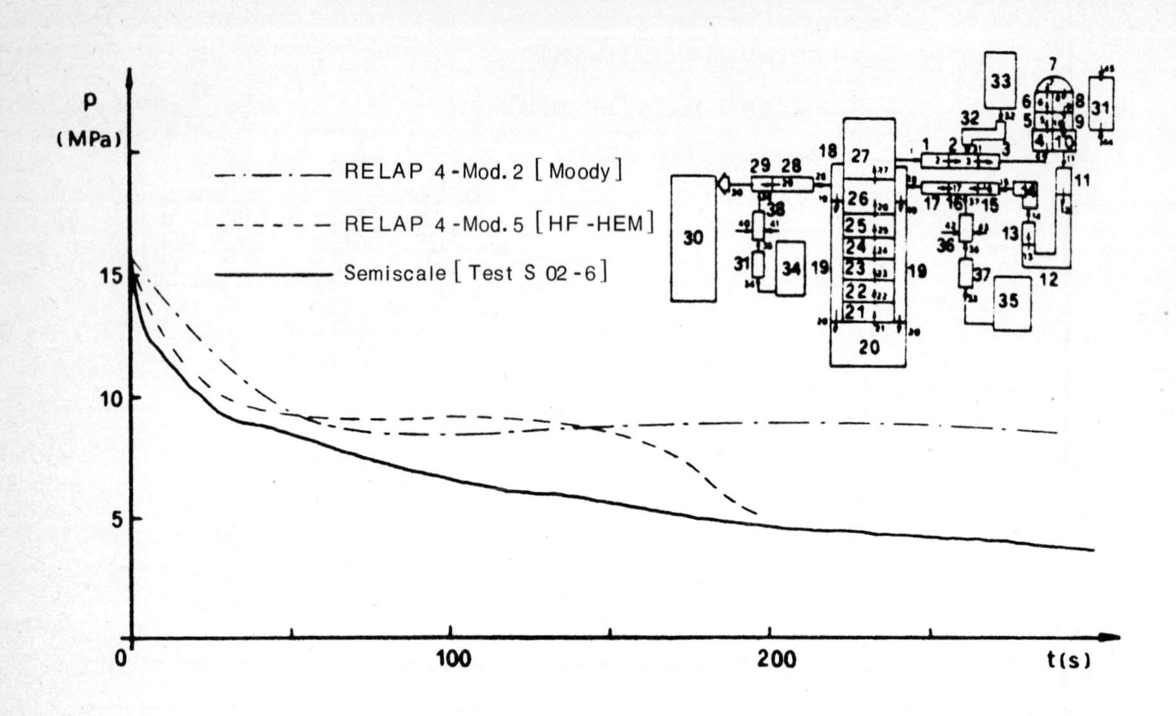

Figure 3-15 **CERULLO ET AL. 1980** /7/
LOWER PLENUM DEPRESSURIZATION

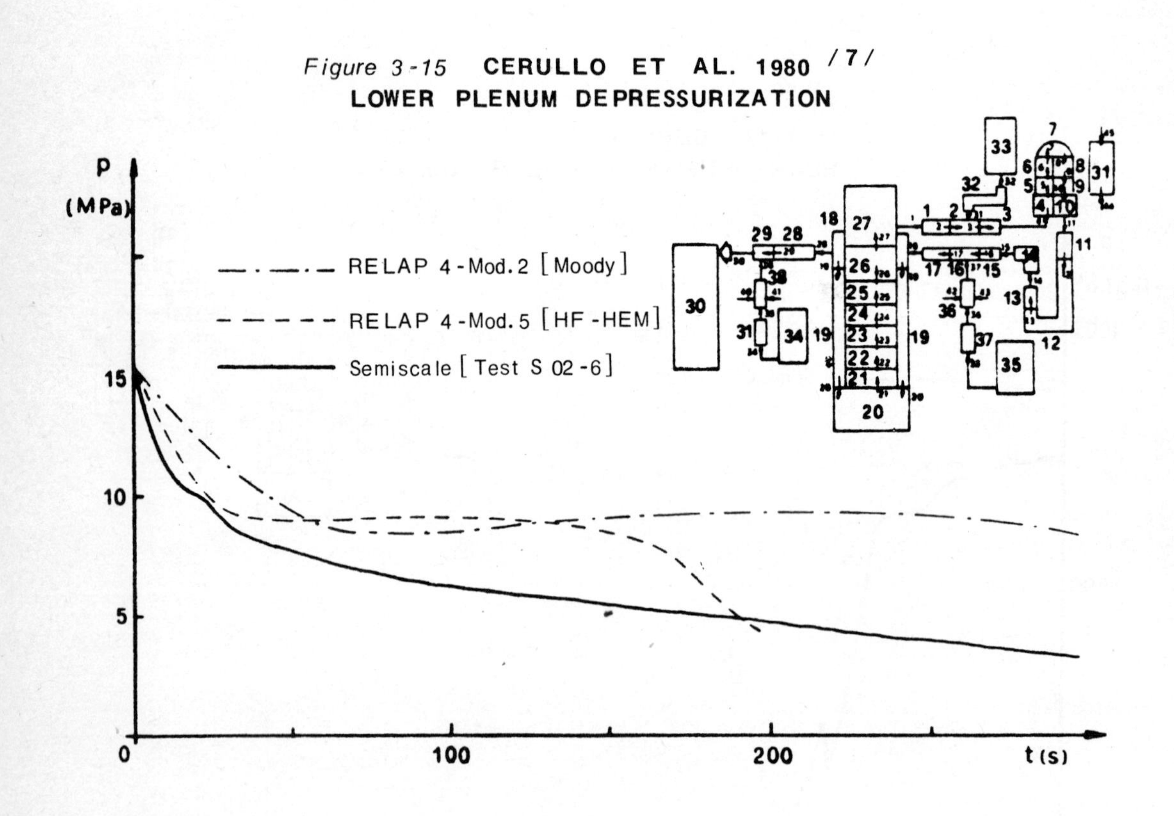

REFERENCES

/1/ F. D'AURIA, P. VIGNI, "Two-phase Critical Flow Models - A Technical Addendum to the CSNI State-of-the-Art Report on Critical Flow Modelling". CSNI Report No. 49, Instituto di Impianti Nucleari, Università di Pisa under work sponsored by the Comitato Nazionale per l'Energia Nucleare, Report Published May 1980.

/2/ A. TENTNER, J. WEISMANN, "The use of the Method of Characteristics for Examination of Two-phase Flow Behaviour", NT, 1978, n. 37.

/3/ G.W. BURNETTE, G.L. SOZZI, "Synopsis of the BWR blowdown heat transfer program", NS, 1979, n. 20.

/4/ RELAP4/MOD5, "A Computer Program for Transient Thermalhydraulic Analysis of Nuclear Reactors and Related Systems", ANCR-Nureg-1335, 1976.

/5/ G.B. WALLIS, "Critical Two-phase Flow", EPRI Workshop, 1979.

/6/ D.G. HALL, "A Study for Critical Flow Prediction for SEMISCALE MOD.1 Loss-of-Coolant Accident Experiments", Tree-Nureg-1006, EG&G, Idaho Inc., 1976.

/7/ N. CERULLO, B. GUERRINI, F. ORIOLO, P. VIGNI, "Thermohydraulic problems in LWRs related to the assessment of LOCA computer RELAP4 code", Instituto di Impianti Nucleari, Università di Pisa, RP 414 (80)

/8/ J.A. BOURE, "The Critical Flow Phenomenon with Reference to Two-Phase Flow and Nuclear Reactor Systems". Reprinted from "Symposium on the Thermal and Hydraulic Aspects of Nuclear Reactor Safety", Volume 1, Ed. O.C. Jones and S.G. Bankoff.

4

CRITICAL FLOW EXPERIMENTAL DATA

Because of its importance, a large number of experimental studies of the phenomenon of critical flow have been conducted during the last 40 years. The necessity of properly understanding critical flow in order to predict the results of a nuclear reactor loss-of-coolant accidents was primarily responsible for accelerating the rate at which experimental studies were conducted beginning in the 1960s. The results of experimental studies have been used to develop numerous analytical models of the phenomenon. It is clear that the problem of predicting the critical flow rate and the thermodynamic state at the choke point is not a closed case for many flow situations of interest. This problem is a result of the large variety of flow geometries of interest and the wide range of fluid conditions over which the phenomenon must be predicted.

In order to assess critical flow models and improve their capabilities, it is necessary to refer to the body of experimental data that is available and suitable. This chapter presents the results of an inventory of the critical flow experimental data base. The principal parameters describing each experimental study that has contributed to the data base are presented in the next section, followed by a discussion of ranges of parameters for which data are available. The fourth section of this chapter contains conclusions regarding the availability of experimental critical flow data and recommendations for future experimental work. The chapter concludes with a bibliography of references describing experimental critical flow studies.

4.1 DATA BASE INVENTORY

The experimental critical flow data base was inventoried by reviewing as many references documenting experimental critical flow studies as possible. Documents were selected from computerised literature searches of the catalogues of the following information services for the indicated years:

Nuclear Safety Information Service - 1967-1980

DOE Energy Data Base - 1974-1980

Nuclear Science Abstracts - 1967-1976

Government Reports Announcements (National Technical Information Service) - 1964-1976

Engineering Index - 1970-1979

Science Abstracts - 1969-1979

References for years prior to 1964 were identified from reference sections and bibliographies contained in the later reports.

The critical flow data base is summarised in Table 4-1. This table lists parameters that characterise each experimental programme. The table does not present a complete inventory of the experimental data base, but does contain many of the experimental data sources that have been referenced in the literature during the past 20 years. The table entries are divided into four groups: pipes, nozzles, orifices, and other geometries. The entries within each group are listed in chronological order from the most recent to the earliest. The author's name appearing on the reference documenting the experiment is used to identify each study in the table.

Exceptions to this convention have been made for the MARVIKEN CFT Project and the Semiscale Project. Data from these programmes are referred to by the project name. The document publication date is given generally by month and year. This date does not necessarily correspond closely to when the testing was conducted. The general type of test section or flow geometry in which choking occurred is listed, followed by the size or range of sizes of the minimum test section cross-section.

Table 4-1

CRITICAL FLOW EXPERIMENTAL DATA BASE

SOURCE	DATE	TYPE	SIZE (mm)	FLUID	REGIME	PRESSURE LEVEL (bars)	COMMENTS
		P	I	P	E	S	
Boivin[1]	12-79	pipe (rec)	12-50	H_2O	sub	p_o = 20-101	L/D = 38 - 53 ; test section lengths not clearly reported
Marviken CFT Project[2]	1978-79	pipe (rec)	200-509	H_2O	sub & sat (2ϕ)	p_o = 40-50	L/D = 0.3 - 3.7
Jeandey and Pinet[3]	6-78	pipe (sec?)	14	H_2O/N_2	simulated 2ϕ	p_p = 2-6	L/D = 169; stagnation conditions not reported; pressure in upstream portion of the pipe (p_p) reported
Ardon and Ackerman[4]	6-78	pipe (sec)	26	H_2O	sub	p_p = 1.4	L/D = 39; stagnation conditions not reported; static pressure in upstream portion of the pipe (pp) reported
Reocreux[5]	8-77	pipe (sec)	20	H_2O	sat (2ϕ)	p_e = 1.5-2.0	L/D = 124; stagnation conditions not reported; static pressure at the exit of the constant section(p_e) reported
Semiscale Project[6]	6-77	pipe (cec)	18	H_2O	sub & sat (2ϕ)	p_o = 3-103	L/D = 4; system blowdown experiment
Rassokhin, et al.[7]	5-77	pipe (sec)	30	H_2O	sub & sat	p_o = 1-32	L/D = 0.3; flow rates not reported
Khlestkin, Kanishchev, and Keller[8]	3-77	pipe (sec)	4	H_2O	sub & sat	p_o = 6-228	L/D = 0.5-6.0; flow rates are in nondimensional form
Prisco, Henry, Hutcherson, and Linehan[9]	3-77	pipe (sec)	20	Freon-11	sub & sat (2ϕ)	p_o = 67-115 kPa	L/D = 2.8 - 100.0
Morrison[10]	10-76	pipe (rec)	28	H_2O	sub & sat (2ϕ)	p_o = 58-67	L/D = 4.8

Note: cec = conical entrance contour
rec = radiused entrance contour
sec = sharp entrance contour

sat = saturated liquid state
sat (2ϕ) = saturated two-phase state
sub = subcooled state

Table 4-1

CRITICAL FLOW EXPERIMENTAL DATA BASE (Cont.)

SOURCE	DATE	TYPE	SIZE (mm)	FLUID	REGIME	PRESSURE LEVEL (bars)	COMMENTS
Seynhaeve, Giot, and Fritte[11]	8-76	pipe (sec & cec)	12.5, 20	H_2O	sub	p_e = 1.4-6.7	L/D = 17.7 - 124.5; stagnation conditions not reported; static pressure at the exit of the constant area section (p_e) reported
Hutcherson[12]	11-75	pipe (cec)	108	H_2O	sat (2ϕ)	p_o = 1-18	L/D = 3; system blowdown experiment
Sozzi & Sutherland[13]	7-75	pipe (sec & rec)	13	H_2O	sub & sat (2ϕ)	p_o = 30-71	L/D = 0.4-140
Prisco[14]	2-75	pipe (sec)	8	CCl_3F	sub?	po = 67-115 Pa	L/D = 2.8-12.8
Howard[15]	1-75	pipe (sec & rec)	2-6	Freon-11	sub	p_o = 52-165 kPa	L/D = 25-300
Edwards & Jones[16]	1974	pipe (sec)	32	H_2O	sat(2ϕ)	p_o = 2-54	L/D = 28; system blowdown experiment
Mal'tsev, Khlestkin, and Keller[17]	6-72	pipe (sec)	3, 3.5	H_2O	sat	p_o = 20-220	L/D = 0.5-9.0
Klingebiel & Moulton[18]	3-71	pipe (cec)	13	H_2O	sat (2ϕ)	p_e = 2-5	L/D = 44; stagnation conditions not reported; static pressure at exit of constant area section (p_e) reported
Henry[19]	9-70	pipe (rec)	8	H_2O	sub	p_e = 10-20	L/D = 115; stagnation conditions not completely reported; static pressure at exit of constant area section (p_e) reported
Allemann et al.[20]	6-70	pipe (sec)	21-173	H_2O	sub & sat (2ϕ)	p_o = 42-165	L/D = 0.5-4.3; system blowdown experiment
Henry[21]	3-68	pipe (rec)	3, 8	H_2O	sub	p_e = 2-10	L/D = 115, 274; stagnation conditions not fully reported; static pressure at exit of constant area section (p_e) reported

Note: cec = conical entrance contour
rec = radiused entrance contour
sec = sharp entrance contour

sat = saturated liquid state
sat (2ϕ) = saturated two-phase state
sub = subcooled state

Table 4-1

CRITICAL FLOW EXPERIMENTAL DATA BASE (Cont.)

SOURCE	DATE	TYPE	SIZE (mm)	FLUID	REGIME	PRESSURE LEVEL (bars)	COMMENTS
Kelly[22]	1-68	pipe (sec?)	2-3	H_2O	sub & sat (2ϕ)	p_e = 1-6	L/D = 90; stagnation conditions not specified; static pressure at exit of constant areas section (p_e) reported
Uchida & Nariai[23]	8-66	pipe (sec)	4	H_2O	sub & sat	p_o = 0.2-0.8	L/D = 25-625
Fauske[24]	1965	pipe (sec)	6	H_2O	sat	p_o = 7-124	L/D = 0-40
Zaloudek[25]	1-64	pipe (cec)	13	H_2O	sat (2ϕ)	p_o = 28-124	L/D = 20
Zaloudek[26]	5-63	pipe (sec & rec)	6-16	H_2O	sub	p_o = 8-25	L/D = 1-6
Cruver[27]	1963	pipe (cec)	13	H_2O	sat (2ϕ)	p_e = 1-3	L/D - 52; stagnation conditions not fully reported; static pressure at the exit of the constant area duct (p_e) reported
Fauske & Min[28]	1-63	pipe (sec?)	?	Freon-11	sat	p_o = 103 kPa	L/D = 2-55
Fauske[29]	10-62	pipe (sec)	3-12	H_2O	sat (2ϕ)	p_e = 3-25	L/D = 228-880; stagnation conditions not reported; static pressure at exit of constant area section ((p_e) reported
James[30]	1962	pipe (?)	76, 152, 203	H_2O	sat (2ϕ)	p_e = 1-4	Test section length not reported; stagnation pressure not reported; static pressure at exit of constant area duct (p_e) reported
Friedrich & Vetter[31]	1-62	pipe (sec & rec)	4	H_2O	sub & sat (2ϕ)	p_o = 6-30	L/D = 0.2-15
Friedrich[32]	10-60	pipe (sec & rec)	1.5-4	H_2O	sub & sat (2ϕ)	p_o = 2-61	L/D = 0.2-2.5

Note: cec = conical entrance contour
rec = radiused entrance contour
sec = sharp entrance contour

sat = saturated liquid state
sat (2ϕ) = saturated two-phase state
sub = subcooled state

Table 4-1

CRITICAL FLOW EXPERIMENTAL DATA BASE (Cont.)

SOURCE	DATE	TYPE	SIZE (mm)	FLUID	REGIME	PRESSURE LEVEL (bars)	COMMENTS
Isbin, Moy, and Da Cruz[33]	9-57	pipe (cec)	10-26	H_2O	sat	p_e = 27-296 kPa	L/D = 23-64 assuming L = 610 mm (2 ft.) stagnation conditions not reported; static pressure at exit of constant area section (p_e) reported
Moy[34]	1-55	pipe (cec)	6-25	H_2O	sat (2ϕ)	p_e = 27-296 kPa	L/D = 35-96; stagnation conditions not reported; static pressure at exit of constant area section (p_e) reported
Pasqua[35]	5-52	pipe (sec & rec)	1-3	Freon-12	sub & sat	6-9	L/D = 4-24
Linning[36]	1952	pipe (sec?)	1.5,3	H_2O	sub	p_o = 2	L/D = 1125, 2400
Burnell[37]	12-47	pipe (sec & rec)	5-38	H_2O	sat	p_o = 1-12	L/D = 0-656
Silver & Mitchell[38]	1945	pipe (rec)	5,13	H_2O	sub & sat	p_o = 1-3	L/D = 0.3-11.4
Danforth[39]	5-41	pipe (rec)	3	H_2O	sub	p_o = 3-7	L/D = 1
			N O Z Z L E S				
Martinec[40]	12-79	Nozzle	3	Freon-11	sub	p_o = 16-22	
Zimmer et al.[41]	4-79	Nozzle	25	H_2O	sub	p_o = 1-10	
Semiscale Project[42]	12-78	Nozzle	17	H_2O	sub & sat (2ϕ)	p_o = 3-100	System blowdown experiment
Karasev, Vazinger, and Mingaleeva[43]	6-77	Nozzle	4,19	H_2O	sat	p_o = 20-100	

Note: cec = conical entrance contour
rec = radiused entrance contour
sec = sharp entrance contour

sat = saturated liquid state
sat (2ϕ) = saturated two-phase state
sub = subcooled state

Table 4-1

CRITICAL FLOW EXPERIMENTAL DATA BASE (Cont)

SOURCE	DATE	TYPE	SIZE (mm)	FLUID	REGIME	PRESSURE LEVEL (bars)	COMMENTS
Semiscale Project[44]	6-77	Nozzle	25	H_2O	sub & sat (2ϕ)	p_o = 3-90	System blowdown experiment
Semiscale Project[45]	1-77	Nozzle	4	H_2O	sub & sat (2ϕ)	p_o = 17-124	System blowdown experiment
Morrison[10]	10-76	Nozzle	28	H_2O	sub & sat (2ϕ)	p_o = 58-67	
Shrock, Starkmann, and Brown[46]	8-76	Nozzle	4-11	H_2O	sub & sat (2ϕ)	p_o = 8-91	
Semiscale Project[47]	7-76	Nozzle	13	H_2O	sub & sat (2ϕ)	p_o = 3-110	System blowdown experiment
Simoneau[48]	12-75	Nozzle	4	N_2	sub	p_o = 5-66	
Semiscale Project[49]	11-75	Nozzle	18	H_2O	sub & sat (2ϕ)	p_o = 6-103	System blowdown experiment
Hendricks, Simoneau, and Barrows[50]	9-75	Nozzle	4	N_2	sub & super-critical	p_o = 9-102	
Sozzi & Sutherland[13]	7-75	Nozzle	13-76	H_2O	sub & sat (2ϕ)	p_o = 30-71	
Dryndrozhik[51]	2-75	Nozzle	6,11	H_2O	sat (2ϕ)	p_o = 2-5	
Adachi & Yamamoto[52]	12-74	Nozzle	10	H_2O	sat (2ϕ)	p_o = 18-30	
Hendricks, Simoneau, and Ehlers[53]	8-72	Nozzle	3	N_2	sub & super-critical	po = 12-102	
Deich et al.[54]	4-69	Nozzle	32.5	H_2O	sat (2ϕ)	p_o = 1	
Vogrin[55]	7-63	Nozzle	5	Air/H_2O	simulated 2ϕ	p_o = 1-7	
Neusen[56]	1-62	Nozzle	6,11	H_2O	sat (2ϕ)	p_o = 8-65	

Note: cec = conical entrance contour
rec = radiused entrance contour
sec = sharp entrance contour

sat = saturated liquid state
sat (2ϕ) = saturated two-phase state
sub = subcooled state

Table 4-1

CRITICAL FLOW EXPERIMENTAL DATA BASE (Cont.)

SOURCE	DATE	TYPE	SIZE (mm)	FLUID	REGIME	PRESSURE LEVEL (bars)	COMMENTS
			ORIFICES				
Covelli[57]	1976	orifice	22.5, 30	sand/water	sat (2ϕ)	2-4	
Edwards & Jones[16]	1974	orifice	22.5	H_2O	sat (2ϕ)	p_o = 2-54	System blowdown experiment
Uchida & Nariai[23]	8-66	orifice	4	H_2O	sub & sat	p_o = 0.2-0.8	
Zaloudek[26]	5-63	orifice	13	H_2O	sub	p_o = 8	
Friedrich & Vetter[31]	1-62	orifice	4	H_2O	sub & sat (2ϕ)	p_o = 6-30	
Friedrich[32]	10-60	orifice	1.5-4	H_2O	sub & sat (2ϕ)	p_o = 2-61	
Monroe[58]	1-57	orifice	6-16	H_2O	sat	p_o = 2-11	
Pasqua[35]	5-52	orifice	1-3	Freon-12	sub & sat	p_o 6-9	
Silver & Mitchell[38]	1945	orifice	5	H_2O	sub & sat	p_o = 1-3	
Benjamin & Miller[59]	7-41	orifice	6-22	H_2O	sat	p_o = 1-21	
			OTHER				
Martinec[40]	12-79	globe valve, relief valve	3 4	Freon-11	sub	p_o = 6-22	
Grison & Lauro[60]	12-78	pump	80 entrance	H_2O	sat (2ϕ)	p_i = 35-85	Stagnation conditions not reported; static pressure at pump inlet (p_i) reported
Zaloudek[61]	3-65	tee, elbow	16	H_2O	sat (2ϕ)	p_e = 1-6	Stagnation conditions not reported; static pressure near exit constant area section (p_e) reported

Note: cec = conical entrance contour
rec = radiused entrance contour
sec = sharp entrance contour

sat = saturated liquid state
sat (2ϕ) = saturated two-phase state
sub = subcooled state

Table 4-1

CRITICAL FLOW EXPERIMENTAL DATA BASE (Cont.)

SOURCE	DATE	TYPE	SIZE (mm)	FLUID	REGIME	PRESSURE LEVEL (bars)	COMMENTS
Fauske & Min[28]	1-63	aperture (9 shapes)	d = 2-7 equivalent	Freon-11	sat	p_o = 103 kPa	
Faletti[62]	12-59	annulus (cec)	d = 5-9 equivalent	H_2O	sat (2ϕ)	p_e = 2-7	Equivalent L/D = 3-107; Stagnation conditions not completely reported; static pressure at exit of constant area section (p_e) reported
Moy[34]	1-55	annulus	d = 6-25 equivalent	H_2O	sat (2ϕ)	p_e = 27-296 kPa	Equivalent L/D = 35-96; stagnation conditions not reported; static pressure at exit of area section (p_e) reported

Note: cec = conical entrance contour
rec = radiused entrance contour
sec = sharp entrance contour

sat = saturated liquid state
sat (2ϕ) = saturated two-phase state
sub = subcooled state

4.2 DISCUSSION OF THE DATA BASE

From Table 4-1, it is clear that the majority of the experimental critical flow data have been obtained using constant area ducts (References /1/ through /39/). A large number of critical flow experiments have been conducted using converging-diverging nozzles (References /10/, /13/, /40/ through /56/). Data for critical flow occurring in orifices are also available (References /16/, /23/, /26/, /31/, /32/, /57/ through /59/), but are quite limited compared to the other two classes of geometry. Only three references documenting studies of critical flow in plumbing components were found: one using tees and elbows (Reference /61/), one using valves (Reference /40/), and one using a pump (Reference /60/). Data on critical flow through geometries resembling a split or crack in a pipe wall or a weldment also seem to be very limited. Only one reference for this type of geometry was found (Reference /28/). It is also noteworthy that little of the pipe and nozzle data were obtained using flow geometries that can be considered ideal from the standpoint of avoiding flow separation at the entry to the constant area section or throat.

The length and diameter of pipe geometries for which references of experimental studies were found are presented in Figure 4-1. Only dimensions of geometries that were tested using water are presented. Differences in entrance contour are denoted by an open (clear) symbol for data from 90 degree entrances, a solid symbol for rounded entrances, and a partially solid symbol for conical entrances. Figure 4-1 shows that critical flow data are available for pipes having diameters ranging from 1.5 to 500 mm, and lengths ranging from 0.6 to 2800 mm. It is clear from this figure that a great deal of data are available for small diameter test sections (less than 13 mm), and only the MARVIKEN CFT Project has provided data for test sections having diameters greater than or equal to 200 mm. In addition, only five experiments with test sections between 30 and 200 mm diameter were found.

Sizes are given to the nearest half millimeter. The test fluid is indicated in Table 4-1. While the majority of the experiments were conducted using water, freon, nitrogen, and gas-water data are also given. The thermodynamic regime(s) in which the flow stagnation conditions resided is listed in the table as "sub", "sat" or "sat(2)", denoting sub-cooled, saturated liquid, and saturated two-phase mixture conditions, respectively. A range of pressures at which data were recorded is included as a table parameter. It was intended that this parameter would refer to stagnation pressure; however, some references did not report this. For these, the pressure measurement nearest the end of the constant area section was considered. Therefore, the range of this pressure has been presented for these references.

All test sections indicated in Table 4-1 containing a constant area section have been designated as "pipes" regardless of size, unless the constant area section was both preceded and followed by a section of varying area. The entrance contour is indicated for each test section that has been designated as a pipe. A 90-degree entrance is indicated by "SEC" (sharp entrance contour), a rounded entrance is indicated by "REC" (round entrance contour), and a conical entrance by "CEC" (conical entrance contour). The exit contour following the constant area duct has not been indicated. Most of the pipes had 90-degree exits. However, some had conical exit contours (e.g. Henry /19/, /21/, Prisco /14/, Réocreux /5/). The term "nozzle" has been used for flow geometries having a varying area section. The entrance or diffuser sections may have been conical or of varying radius. The nozzle throat may have been a single cross-section or a short constant area section. The term "orifice" has been used to denote flow geometries having a 90-degree entrance and a constant area section having an L/D of 0.1 or less.

Several studies have been conducted using test sections having the same diameter, but differing in length over a wide range (References /13/, /23/, /24/ and /38/). Data are also available for cases in which the test section length was held constant and the diameter was varied (References /20/, /29/ and /33/), although the range of diameter variation was generally more limited than the variation in length at constant diameter for these tests.

In order to illustrate the availability of critical flow data in pipes of constant length-to-diameter ratio (L/D), lines of constant L/D have been added to data presented in Figure 4-1 and are shown in Figure 4-2. This figure shows that data are available for L/Ds ranging from less than 1 to over 500. Figure 4-2 shows that data produced using test sections covering a wide range of size are available at the same L/D, for L/Ds less than four. However, comparing experimental results at the same L/D is hampered by differences in entrance contour. Another factor complicating the comparison of data at the same L/D is that the data are seldom available at the same stagnation conditions.

The throat sizes of converging-diverging nozzles for which experimental critical flow data were found are presented in Figure 4-3. Throat sizes vary from 4 to 75 mm. The data show that there is little repetition of throat size. It should be noted that the nozzles vary in entrance contour (conical versus rounded) and in the extent of the minimum area section (a single axial location versus short constant area section).

The sizes of orifices for which experimental critical flow data was found are presented in Figure 4-4. The range of sizes (4 to 30 mm except for the two large diameter Japanese tests) was quite limited compared to the other two classes of geometries. Figure 4-4 shows that only half the orifice sizes have been used in more than one experimental study.

4.3 CONCLUSIONS AND RECOMMENDATIONS

The conclusion that can be drawn from this review of the critical flow data inventory are presented in this section. Recommendations for remedying deficiencies in the data base and for improving the design and reporting of future experimental programs are given.

4.3.1 CONCLUSIONS

- A large amount of experimental critical flow data are available, covering flow in constant area ducts, converging-diverging nozzles, orifices, pipe tees and elbows, valves, and slits.

- More than half of the references found documented critical flow studies conducted with constant area ducts. The test sections covered large ranges of diameter (1.5 to 500 mm) and length (0.6 to 2800 mm).

- Significant amounts of critical flow data are available for converging-diverging nozzles and orifices over a limited range of sizes. The nozzle throat and orifice diameters ranged from 4 to 75 mm and 4 to 125 mm, respectively.

- Little data are available for critical flow in standard piping components and for geometries resembling piping failures other than a guillotine break. Three references were found in which critical flow was studied in standard plumbing components (small scale elbows, tees, valves, and a small scale pump). Only one reference presented critical flow data in slits simulating a pipe failure and again the apparatus used was small scale.

- The 27 tests conducted during the MARVIKEN CFT Project are the only known source of data obtained at high pressure/temperature conditions using test sections greater than 200 mm diameter. Large amounts of data are available only for diameters less than 13 mm, but little for the range 20 to about 100 mm.

- Little data are available for idealised flow geometries designed to avoid entrance separation. Most of the constant area ducts had 90-degree, conical, entrances of small radius (i.e. approximately half the test section diameter). Most nozzles had conical entrances of large half-angle and many had an abrupt change in slope at the entrance of the minimum area section and large half-angle diffusers. Very few test sections had gradual approaches to the minimum area section with a continuous change in slope.

- Differences in test section entrance contour, nozzle throat geometry, and diffuser angle contribute additional uncertainties if the data are used to assess the effect of geometric variables. Such assessments are complicated by a lack of data at common stagnation conditions. Several references did not contain sufficient data to specify completely the stagnation conditions in the fluid, which greatly limits their usefulness for critical flow model assessment and development.

4.3.2 RECOMMENDATIONS

- The existing data base should be carefully reviewed as part of planning for future critical flow research to expand the data base and avoid unnecessary duplication of effort. Test section geometries and stagnation fluid conditions should also be selected to ensure that straightforward comparisons can be made with existing data.

- The inclusion of tabulated data in reporting an experimental critical flow study is recommended as this greatly increases the usefulness of the data. This practice eliminates the need for taking data from report figures. Reporting of sufficient data to define completely the stagnation thermodynamic conditions of the fluid is important because the flow rate and critical thermodynamic state are primarily a function of the stagnation conditions. Furthermore, most critical flow models require the stagnation conditions as input to compute the critical flow rate and critical state. Experimental data which does not include a complete definition of the stagnation conditions thus cannot be used for model assessment.

- Additional data on critical flow in plumbing components and failure geometries such as splits and cracked welds is needed. Since the Three Mile Island Accident, it has become apparent that increased emphasis must be placed on breaks having higher probabilities of occurrence that the guillotine break.

- The range of sizes of converging-diverging nozzles and orifices for which critical flow data is available is rather limited and may need to be expanded. Most data for convering-diverging nozzles and orifices have been obtained with test sections having minimum area sections ranging in size from 4 to 20 or 30 mm. Additional data may be required if larger scale applications of these devices are identifed, such as large scale metering devices.

- More complete information on the uncertainties of all measurements made in a critical flow study are needed, with particular emphasis on the uncertainty of the measured flow rate. Knowing the uncertainty in the data is essential for model or code assessment especially to determine if parametric influences are significant in causing data spread. Less than half the references contained uncertainty information and of what was reported flow measurement uncertainty was generally missing.

- Critical flow data exists for constant area ducts ranging from 1.5 to 500 mm diameter. This suggests that the effects of scale could be assessed. This is, however, hampered by variations in entrance contours and L/D ratios. Further testing should be undertaken to fill in the data set in order to permit assessing scale effects.

- The utility of the existing experimental data would be greatly increased if they were available in a uniform format with supporting software for rapid retrieval and data display and manipulation. Consideration should be given to assembling the data identified in this study into a topical data bank.

Figure 4-1 **LENGTHS AND DIAMETERS OF CONSTANT AREA TEST SECTIONS**

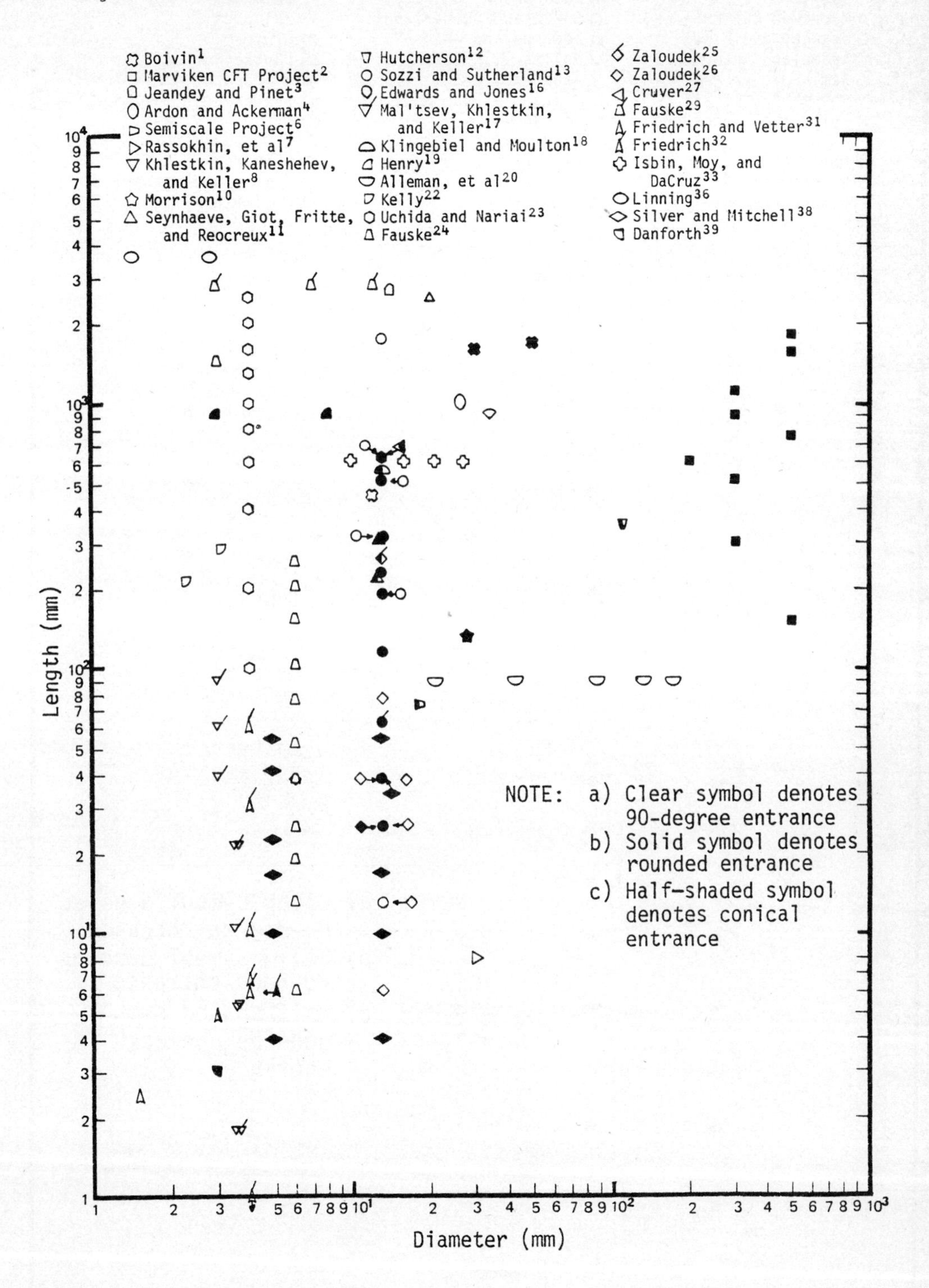

Figure 4-2 **LENGTH-TO-DIAMETER RATIO OF CONSTANT AREA TEST SECTIONS**

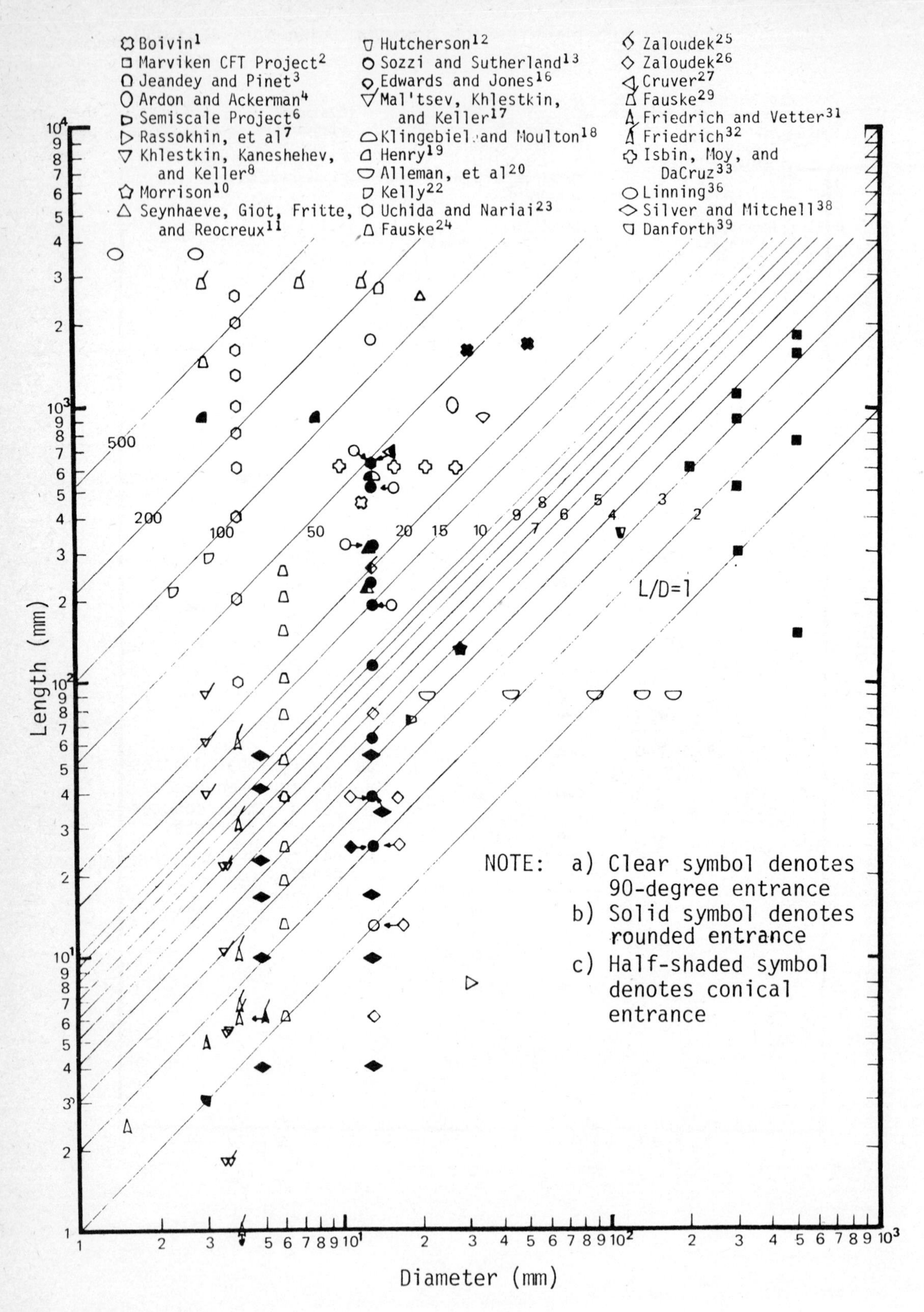

Figure 4-3 **CONVERGING - DIVERGING NOZZLE THROAT DIAMETERS**

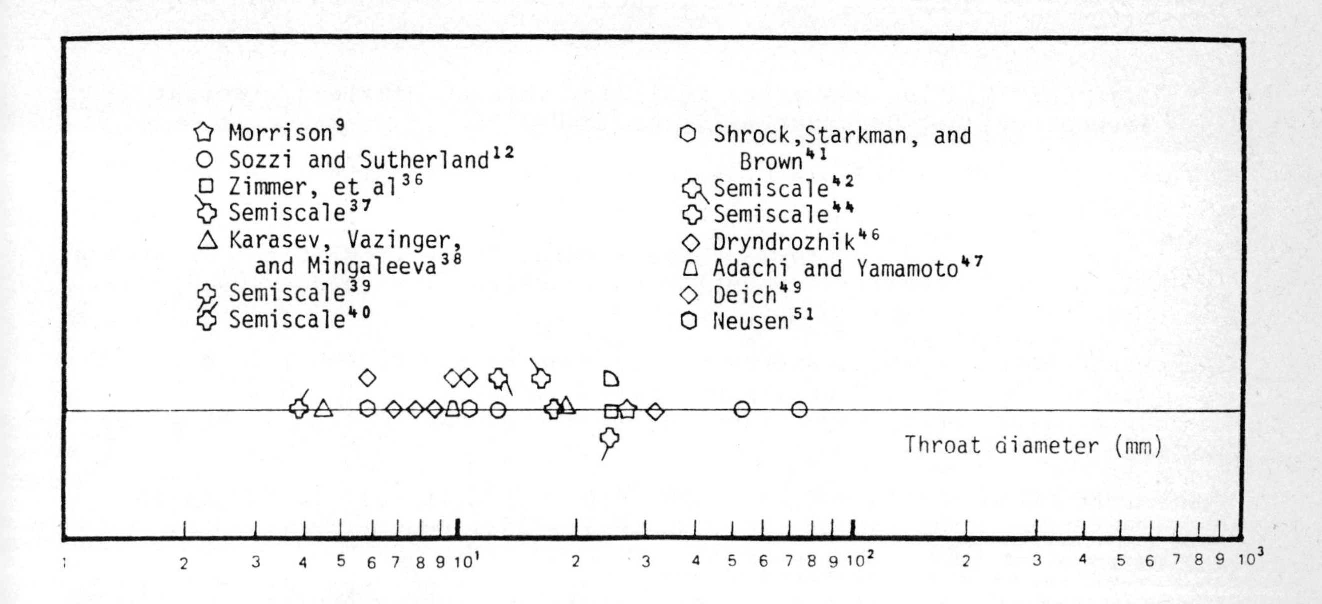

Figure 4-4 **ORIFICE DIAMETERS**

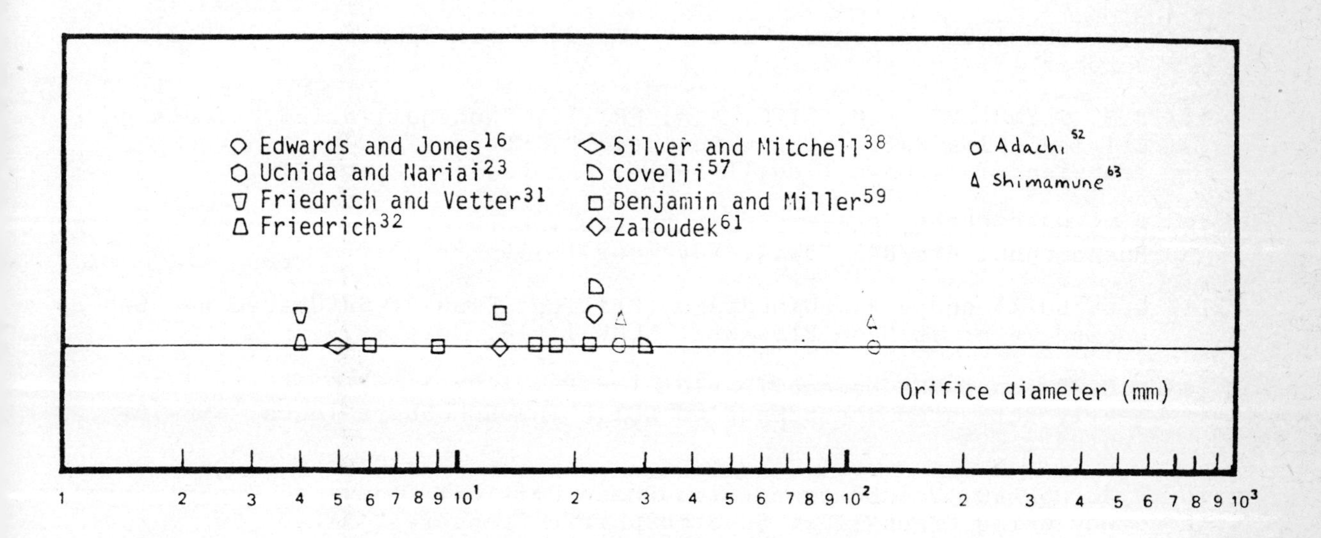

REFERENCES

/1/ J.Y. BOIVIN, "Two-Phase Critical Flow in Long Nozzles", Nuclear Technology, 46, December 1979, pp. 540-545.

/2/ The MARVIKEN Full-Scale Critical Flow Tests; Results of Tests 1-27, MXC201-227.

/3/ C.H. JEANDEY and B. PINET, "Experimental Study of Critical Two-Phase Flow", OECD Specialists Meeting on Transient Two-Phase Flow, Paris, France, 12-14 June 1978.

/4/ K.H. ARDRON and M.C. ACKERMAN, "Studies of the Critical Flow of Subcooled Water in a Pipe", paper presented at the OECD/CSNI Specialists Meeting on Transient Two-Phase Flow, Paris, France, 12-14 June 1978.

/5/ M. REOCREUX, Contribution to the Study of Critical Flow Rates in Two-Phase Water Vapour Flow, NUREG-TR-0002, August 1977.

/6/ V. ESPARZA and K.E. SACKETT, Experimental Data Report for Semiscale Mod-1 Test S-06-5 (LOFT Counterpart Test), TREE-NUREG-1125, June 1977.

/7/ N.G. RASSOKHIN, "Critical Conditions with Unsteady-Stage Outflow of a Two-Phase Medium with a Pipe Break", High Temp, 15, 3, 1977, pp. 491-497.

/8/ D.A. KHLESTKIN, V.P. KANISHCHEV, V.D. KELLER, "Flow Characteristics for Hot Water at an Initial Pressure of 22.8 MPa Escaping into the Atmosphere", Soviet Atomic Energy, 14, 3, 1977, pp. 239-241.

/9/ M.R. PRISCO et al., "Nonequilibrium Discharge of Saturated and Subcooled Liquid Freon-11", Nuclear Science and Engineering, 63, 1977, pp. 365-375.

/10/ A.F. MORRISON, Blowdown Flow in the BWR Test Apparatus, GEAP-21656, October 1977.

/11/ J.M. SEYNHAEVE, M.M. GITO, A.A. FRITTE, "Nonequilibrium Effects on Critical Flow Rates at Low-Qualities", Specialists Meeting on Transient Two-Phase flow, Toronto, Canada, 3-4 August 1976.

/12/ M.N. HUTCHERSON, Contribtuion to the Theory of the Two-Phase Blowdown Phenomenon, ANL/RAS 75-42, November 1975.

/13/ G.L. SOZZI and W.A. SUTHERLAND, Critical Flow of Saturated and Subcooled Water at High Pressure, NEDO-13418, July 1975.

/14/ M.R. Prisco, The Nonequilibrium Two-Phase Critical Discharge of Nearly Saturated and Subcooled CCl_3F Through Short Tubes, ANL-75-9, February 1975.

/15/ P.A. HOWARD, One Component Two-Phase Critical Flow: An Experimental Study Using Freon-11 at Subatmospheric Pressures, ANL-75-8, January 1975.

/16/ A.R. EDWARDS and C. JONES, An Analysis of Phase IIA Blowdown Tests - The Discharge of High Enthalpy Water from a Sample Volume into a Containment Volume, SRD R.27, 1974.

/17/ B.K. MAL'TSEV, D.A. KHLESTKIN, V.D. KELLER, "Experimental Investigation of the Discharge of Saturated and Subcooled Water at High Pressures", Thermal Engineering, 19, 6, 1972, pp. 85-88.

/18/ W.J. KLINGEBIEL and R.W. MOULTON, "Analysis of Flow Choking of Two-Phase, One Component Mixtures", AIChE Journal, 17, 2, 1971, pp. 383-390.

/19/ R.E. HENRY, An Experimental Study of Low Quality Steam-Water Critical Flow at Moderate Pressure, ANL-7740, September 1970.

/20/ R.T. ALLEMAN et al., Experimental High Enthalpy Water Blowdown from a Simple Vessel Through a Bottom Outlet, BNWL-1411, June 1971.

/21/ R.E. HENRY, A Study of One- and Two-Component, Two-Phase Critical Flows at Low-Qualities, ANL-7430, March 1968.

/22/ J.T. KELLY, Two-Phase Critical Flow, MS Thesis, Massachusettes Institute of Technology, January 1968.

/23/ H. UCHIDA and H. NARIAI, "Discharge of Saturated Water Through Pipes and Orifices", Proceedings of the Third International Heat Transfer Conference, Chicago, Illinois, 7-16 August 1966, pp. 1-12.

/24/ H.R. FAUSKE, "The Discharge of Saturated Water Through Tubes", Chemical Engineering Program Symposium Series, 61, 1965, pp. 210-216.

/25/ F.R. ZALOUDEK, Steam-Water Critical Flow from High Pressure Systems Interim Report, HW-80535, January 1964.

/26/ F.R. ZALOUDEK, The Critical Flow of Hot Water Through Short Tubes, HW-77594, May 1963.

/27/ J.E. CRUVER, Metastable Critical Flow of Steam Water Mixtures, PhD. Thesis, University of Washington, 1963.

/28/ H.K. FAUSKE and T.C. MIN, A Study of the Flow of Saturated Freon-11 Through Apertures and Short Tubes, ANL-6667, January 1963.

/29/ H.K. FAUSKE, Contribution to the Theory of Two-Phase, One-Component Critical Flow, ANL-6633, October 1962.

/30/ R. JAMES, "Steam Water Critical Flow Through Pipes", Proceedings of the Institution of Mechanical Engineers, Thermodynamics and Fluid Mechanics Group, 176, 26, 1962, pp. 741-748.

/31/ H. FRIEDRICH and G. VETTER, "The Influence of Nozzle Shapes on Nozzle Flow Behaviour for Water at Different Thermodynamic States", Energie, 14, 1, 1957, pp. 373-377.

/32/ H. FRIEDRICH, "Flow Through Single-State Nozzles with Different Thermodynamic States", Energie, 12, 1, 1960, pp. 411-419.

/33/ H.S. ISBIN, J.E. MOY, A.J.R. DA CRUZ, "Two-Phase Steam Water Critical Flow", AIChE Journal, 3, 3, 1957, pp. 361-365.

/34/ J.E. MOY, Critical Discharges of Steam-Water Mixtures, MS Thesis, University of Minnesota, January 1955.

/35/ P.F. PASQUA, Metastable Liquid Flow Through Short Tubes, Ph.D Thesis, Northwestern University, May 1952

/36/ D.L. LINNING, "The Adiabatic Flow of Evaporating Fluids in Pipes of Uniform Bore", Institution of Mechanical Engineers, Proceedings Series B, 18, 2, 1952, pp. 64-72.

/37/ J.G. BURNELL, "Flow of Boiling Water Through Nozzles, Orifices, and Pipes", Engineering, 12 December 1947, pp. 572-576.

/38/ R.S. SILVER and A. MITCHELL, "The Discharge of Saturated Water Through Nozzles", North East Coast Institution of Engineers and Shipbuilders, Transactions, 42, 2, 1945, pp. 51-66.

/39/ J.L. Danforth, Flow of Hot Water Through a Rounded Orifice, BS/MS Thesis, Massachusetts Institute of Technology, May 1941.

/40/ E.J. MARTINEC, Two-Phase Critical Flow of Saturated and Subcooled Liquids Through Valves, ANL/RAS/LWR 79-7, December 1979.

/41/ G.A. ZIMMER et al., Pressure and Void Distribution in a Converging-Diverging Nozzle with Nonequilibrium Water Vapor Generation, BNL-NUREG 26003, April 1979.

/42/ R.L. GILLINS, K.E. SACKETT, K. STANGER, Experimental Data Report for Semiscale Mod-3 Blowdown Heat Transfer Test S-07-3 (Baseline Test Series) NUREG/CR-0356, TREE-1223, December 1978.

/43/ E.K. KARASEV et al., "Investigation of the Adiabatic Expansion of Water Vapor from the Saturation Line in Lavel Nozzles", Soviet Atomic Energy, 42, 6, 1977, pp. 545-549.

/44/ E.M. FELDMAN and K.E. SACKETT, Experimental Data Report for Semiscale Mod-1 Tests S-05-6 and S-05-7 (Alternate ECC Injection Tests), TREE-NUREG-1055, June 1977.

/45/ B.L. COLLINS, H.S. CRAPO and K.E. SACKETT, Experimental Data Report for Semiscale Mod-1 Test S-02-6 (Blowdown Heat Transfer Test), TREE-NUREG-1037, January 1977.

/46/ V.E. SCHROCK, E.S. STARKMAN, R.A. BROWN, "Flashing Flow of Initially Subcooled Water in Convergent-Divergent Nozzles" ASME-AICE Heat Transfer Conference, St. Louis, MO., 9-11 August 1976, 76-HT-12.

/47/ H.S. CRAPO, M.F. JENSEN, K.E. SACKETT, Experimental Data Report for Semiscale Mod-1 Test S-29-1 (Integral Test With Symmetrical Break) ANCR-NUREG-1327, July 1976.

/48/ R.J. SIMONEAU, Pressure Distribution in a Converging-Diverging Nozzle during Two-Phase Choked Flow of Subcooled Nitrogen, NASA TMX-71762, 1975.

/49/ H.S. CRAPO, M.F. JENSEN, K.E. SACKETT, Experimental Data Report for Semiscale Mod-1 Test S-02-4 (Blowdown Heat Transfer Test), ANCR-1234, November 1975.

/50/ R.C. HENDRICKS, R.J. SIMONEAU, R.F. BARROWS, "Critical Flow and Pressure Ratio Data for LOX Flowing Through Nozzles", Fourteenth International Congress of the International Institute of Refrigeration, Moscow, USSR, 20-30 September 1975 (NASA TMX-71725; E-8347).

/51/ E.I. DRYNDROZHIK, "Critical Flow Regime of a Low-Quality Steam-Liquid Mixture in Convergent Nozzles", Fluid Mechanics, 4, 1, 1975, pp. 139-144.

/52/ H. ADACHI and N. YAMAMOTO, "High Speed Two-Phase Flow (II), Flashing Flow Through a Converging-Diverging Nozzle", Heat Transfer, 3, 4, 1974, pp. 89-102.

/53/ R.C. HENDRICKS, R.J. SIMONEAU, R.C. EHLERS, "Choked Flow of Fluid Nitrogen with Emphasis on the Thermodynamic Critical Region", Advances in Cryogenic Engineering, 18, 1978, pp. 150-161.

/54/ M.E. DEICH et al., "Investigation of the Flow of Wet Steam in Axisymmetric Laval Nozzles over a Wide Range of Moisture Content", Teplofizika Vysokikh Temperatur, 7, March-April 1969, pp. 327-333.

/55/ J.A. Vogrin Jr., An Experimental Investigation of Two-Phase, Two-Component Flow in a Horizontal, Converging-Diverging Nozzle, ANL-6754, July 1963.

/56/ K.F. NEUSEN, Optimising of Flow Parameters for the Expansion of Very Low-Quality Steam, UCRL-6152, January 1962.

/57/ B. COVELLI, "Critical Flow in Orifices of a Boiling Three-Phase Mixture", Proceedings of the Fifth International Symposium on Fresh Water from the Sea, 2, 1976, pp. 223-232.

/58/ E.S. Monroe, "Flow of Saturated Boiler Water Through Knife Edge Orifices in Series", Transactions of the ASME, 78, 2, 1956, pp. 373-377.

/59/ M.W. BENJAMIN and J.G. MILLER, "The Flow of Saturated Water Through Throttling Orifices", Transactions of the ASME, July 1941, pp. 419-429.

/60/ P. GRISON and J.F. LAURO, "Two-Phase Performance Characteristics of a PWR Primary Pump Under LOCA Conditions", Annual Meeting of the ASME, San Francisco, CA, 10-15 December 1978, pp. 197-212.

/61/ F.R. ZALOUDEK, The Low Pressure Critical Discharge of Steam-Water Mixtures from Pipe Elbows and Tees, BNWL-34, March 1965.

/62/ D.W. FALETTI, Two-Phase Critical Flow of Steam-Water Mixtures, Ph.D Thesis, University of Washington, 1959.

/63/ H. SHIMAMUNE et al., Thermo-hydraulic Behaviour in a Primary Cooling System during a Loss-of-Coolant Accident of a Light Water Reactor (Result of the Mock-up Test with ROSA-1), JAERI-M-6318, 1975.

5

APPLICATION OF CRITICAL FLOW MODELS IN SAFETY CALCULATIONS

5.1 INTRODUCTORY REMARKS

After the inventory of the analytical models for describing and predicting critical flow examined in this report (Chapter 3) and the list of experiments conducted to study this phenomenon (Chapter 4), the next logical step is to compare the models with the experiments. A detailed study of such comparisons, however, reveals the following:

- Almost all the models are said by their authors to give good prediction of experimental results, ranging from "good agreement" to "excellent agreement", sometimes with models that are in total conflict. Actually, the agreement is confined to a small number of parameters and is often based on a limited number of specific tests.

- Any valid comparison between models and data would require independent verification, a very time-consuming task far exceeding the scope of this report.

The two main difficulties are as follows:

- Certain data are difficult to process, due to: insufficient documentation (as stated in Chapter 4); results presented in the form of difficult to use curves; inlet conditions not specified; inlet geometries too vague; and so on.

- The number of models is very large (see Chapter 3) and methods of application are not always explained. For example, some models use reservoir conditions without specifying what should be done when these conditions are not present during testing (e.g. system experiments or actual reactor testing).

For these reasons, this chapter is confined to models currently used in safety calculations. First, we shall list the models found in a number of codes. Physical features of the models and numerical applications will then be compared. Finally, some general conclusions will be given about the models' suitability for predicting experimental results. From this discussion, several recommendations will be made concerning physical and numerical features, confirmatory testing, and practical use of the models.

5.2 LIST OF MODELS USED IN SAFETY CALCULATIONS

The list below does not cover all accident codes. The codes selected represent either the most frequently used or the most representative in terms of critical flow modelling. Rather than making an exhaustive list, we have selected typical examples in order to illustrate the discussion in Section 3, which does cover all codes.

The models are studied from two views:

- Physical modelling, that is, the basic principles on which the analytical models are based. The models are not discussed in detail as the derivation of the equations is fully described in various papers and manuals.

- Numerical "translation" of the models. The calculations used for solving the analytical equations involve numerical methods based on discretised treatment of the equations. It is essential to know how the critical flow models are introduced in the discretised systems, as this may lead to difficulties and even major distortions.

5.2.1 RELAP4 CODE (RELAP4/MOD5 User's Manual, 1976)

5.2.1.1 Physical Models

There are four physical models:

- the "sonic" model,

- the homogenous equilibrium model (HEM) (Hutcherson /1/),

- the MOODY /2/ model,

- the HENRY-FAUSKE /3/ model.

Sonic and homogeneous equilibrium (HEM) models

The sonic and HEM models assume that the two phases are in thermodynamic and mechanical (no slip) equilibrium. The two-phase mixture is considered to be a compressible single-phase fluid, whose average density is a function of the qualities and the densities of each phase. Single-phase flow data can be applied to this fluid. Sound velocity can be calculated by the well-known formula:

$$a^2 = \left(\frac{\partial p}{\partial \rho_m}\right)_S$$

Critical flow is given by:

$$G = \rho_m \, aA$$

The sonic model and the HEM differ in their treatment of fluid changes up to the point at which critical conditions are reached. The sonic model assumes that sound velocity remains contstant, equal to the velocity of the upstream fluid volume under the existing thermohydraulic conditions. The fluid properties, however, are assumed to undergo the usual changes due to drag and changes in kinetic energy between the centre of the volume and the critical section. This is followed by isentropic expansion in the critical section, bringing the fluid to a sonic regime.

The HEM assumes isentropic expansion occurs between the centre of the upstream volume and the critical section. This is similar to the approach used for the isentropic flow of a perfect gas.

MOODY Model

This model assumes the flow to be in thermodynamic equilibrium, but with the phases flowing at different rates. The fluid is assumed to undergo isentropic change starting from stagnation conditions. Thus, at a given stagnation regime the flow can be expressed as a function $G(p,\gamma)$ of the local values of pressure p and slip γ in the fluid at the assumed flow. Critical conditions are defined as the values of pressure and slip maximising the function $G(p,\gamma)$. These values apply solely to stagnation conditions, so flow can be expressed as a function of only these conditions. It is assumed that the critical values of pressure and slip are reached simultaneously in the critical section, through "independent" and "ad hoc" changes in the two parameters, since no mechanism fully accounts for this simultaneity.

HENRY-FAUSKE Model

The HENRY-FAUSKE model uses a large number of assumptions and approximations. As implemented in RELAP, it is assumed that the fluid changes between stagnation conditions and the critical section as follows:

- isentropic flow ,

- slip equal to 1,

- no mass and heat transfer between phases,

- incompressible liquid.

As a result, each phase retains the same mass fraction and entropy while the fluid changes proceed.

By combining the equations of conservation of mass and momentum, the flow can be expressed as a function of the other fluid parameters. Flow is critical if the derivative of this expression with respect to pressure in the critical section is zero. In the resultant expression for critical flow, non-derivative terms are deduced from the fluid change assumptions listed above (e.g. current quality equal to stagnation quality, slip equal to 1). The derivative terms, however, are evaluated on the basis of different assumptions:

- Steam is assumed to expand in polytropic fashion, whereas it would expand isentropically during the fluid's evolution.

- The derivative of the liquid/steam ratio with respect to pressure is calculated as a function proportional to the derivative of the equilibrium liquid/steam ratio with respect to pressure. The proportionality constant M is an essential feature of the model. If the assumption made for the changes in the fluid properties up to the critical section had been used, this derivative would have been zero (constant ratio).

From these assumptions, critical flow can be expressed as a function of stagnation conditions and critical pressure. The latter is removed by combining the flow expression with the momentum equation integrated between stagnation and critical conditions. The flow is then solely a function of stagnation conditions.

Contraction Coefficients

When discussing RELAP4 critical flow models, the well-known "contraction coefficient" must be mentioned. The coefficient enables the value of the flow given by the model to be multiplied by a value usually lower than 1. A tentative explanation on physical grounds is that the critical section is not the total geometric section owing to a "vena contracta" phenomenon. However, this cannot be so because these coefficients are often used even for straight cylindrical pipes and their values must differ from model to model to match test data. For example, MO DY with a coefficient of 0.6 produces results similar to HEM with a coefficient of about 1.0. In fact, these coefficients are merely adjustable constants not explicitly included in the models but frequently applied by users. Indeed, certain values can even be routinely used for a specific model.

5.2.1.2 Numerical Calculation in RELAP4

RELAP4 numerically calculates flow from a representation in terms of volumes and junctions (see Figure 5.1). Continuity and energy equations are integrated over volumes v_n and v_{n+1}, whereas the momentum equation is integrated over a shifted volume (shown in dotted outline, in Figure 5.1). The first two equations give the state of the fluid at the centre of the volumes (pressure, enthalpy, and so on) while the last gives flow at junction J_n. Generally, the first step in the critical flow evaluation is to calculate flow at junction J_n by means of the usual momentum equation. This value is compared with that of the model used.

Critical flow models can be numerically applied in different ways. In the sonic model, sound velocity is calculated using the thermodynamic properties of the fluid at the centre of volume V_n. It is assumed that this sound velocity is the same at junction J_n. The density at J_n is deduced from that calculated at the centre of volume V_n by calculating the changes in the fluid due to drag and heat transfer between V_n and J_n. Added to this is isentropic expansion up to sonic conditions at junction J_n.

In view of the entropy changes between V_n and J_n, sound velocity varies. The method used, therefore, contains an approximation which should be considered when mesh size is being selected. Also, when critical flow is compared to flow calculated by means of the momentum equation, fluid changes between V_n and J_n differ because isentropic expansion is added to the usual changes between V_n and J_n when calculating critical flow.

The HEM, MOODY and HENRY-FAUSKE models are based on the use of stagnation conditions, which are either approximated through static conditions or calculated from the latter. If they are approximated, the error will increase with fluid velocity, which generally increases near the critical section. The error will also vary considerably with mesh size since the areas under investigation are those subject to strong acceleration. These conditions are not considered at the same point, pressure being considered at the centre of volume V_n and enthalpy at J_n. This leads to further uncertainties.

To determine stagnation conditions, stagnation enthalpy at J_n is first calculated by bringing the fluid to rest isentropically from its velocity at the junction. Stagnation pressure is then calculated from the pressure at the centre of volume V_n following isentropic change using the same variation (enthalpy-stagnation enthalpy) as at J_n. Any distortions in these stagnation conditions compared to those calculated from conditions at the centre of the volume depend on the extent of the irreversibilities and the variation in kinetic energy between V_n and J_n.

From the stagnation conditions calculated, the HEM, MOODY and HENRY-FAUSKE models are applied with their specific treatment of fluid properties leading up to critical conditions. Finally, a number of numerical transitions are provided for the smooth transfer from one model to the next.

5.2.2 RELAP5 CODE

5.2.2.1 Physical Model

The critical flow model /4/ in RELAP5 is derived from the flow model. This is a one-dimension transient model using five equations:

- 2 equations of phase continuity,
- 2 equations of phase momentum,
- 1 energy equation for the mixture.

The system obtained consists of partial differential equations as follows:

$$A\,X'_t + B\,X'_z = C$$

Five primary parameters are calculated: velocity of each phase; pressure, assumed to be the same for both phases; and void fraction and temperature of one of the two phases, the other phase assumed to be at saturation. The changes in the five parameters are obtained by specifying the transfer laws between phases and between the phases and the pipe wall. The algebraic parts of the laws are contained in the term C. The authors of the code have also introduced derivatives in these laws, and these occur in matrices A and B.

The critical condition is obtained by applying the theory of characteristics, that is, when one of the roots of the equation with the characteristics det $|A\lambda - B| = 0$ becomes zero with the other roots being positive (i.e. in practice when det $|B| = 0$). The derivative terms of the transfer laws then influence the critical conditions, as shown by Boure /5/ and Reocreux /6/.

In RELAP5 the following are introduced:

- A virtual mass term in the momentum transfer between phases.
- A mass transfer model based on the assumption of thermodynamic equilibrium.

This mass transfer model is different from that used for calculating the flow itself, which is a relaxation-type model giving fluid changes on a thermodynamic non-equilibrium model.

Because this type of mass transfer model is used for determining critical flow, sound velocities show a discontinuity as soon as steam appears (which is familiar to users of the HEM model). This arises for subcooled fluid flow for example, and requires special treatment. Two situations are considered:

- Steam appears at the critical section (pressure equal to p_{sat} or p_{sat-p} to express delayed boiling). Flow is then calculated by means of the Bernoulli equation in the upstream single-phase area (the downstream two-phase area is then under the supersonic regime as defined in the model).

- Steam appears upstream from the critical section, and the previous two-phase model is applied.

Flow values are calculated for both situations and the higher value is selected.

Finally, a model /7/ has been developed in the MOD1 version to handle the case of small breaks in a large pipe with laminar flow. It can be used for calculating the properties of the fluid entering the break as a function of the size of the latter and the respective locations of the break and the level in the main pipe.

The various critical conditions described above are associated with the flow model determining the changes in the fluid properties up to the critical section. Flow characteristics in the critical section will, therefore, be directly given by the flow model and will thus chiefly depend on the laws governing inter-phase mass and momentum transfer used in this model.

5.2.2.2 Numerical Calculation in RELAP5

RELAP5 utilises a discretisation method with offset meshes, the first set of meshes being used for calculating scalar values (at the centres of volumes) and the second offset by half-a-mesh for calculating vector values at the junctions (see Figure 5.2).

The break is placed over a junction, so that the last scalar mesh is shortened by half a mesh. In the numerical calculation half a mesh A_{out} must be added (see Figure 5.2) where the pressure outside the circuit is specified and velocities are assumed to be those calculated at the last junction J_{out}.

This is the framework in which the previoius critical conditions are applied, but first the condition defined above, namely λ_i (or that of det $|B| = 0$) equals zero, is markedly altered into a much easier form to handle. There are no grounds for this change other than it yields "good results" /7/. The new critical equation ($E_c = 0$) is then regarded as the "RELAP5" critical condition.

At each junction and each time step, a check is made to see whether this equation is true by explicitly calculating the terms in the equation ($E_c = 0$). If there is no change in the sign of E_c, there is no critical flow and the calculation proceeds under the conditions within the usual limits. If the calculation shows that the sign of E_c has changed, flow is regarded as critical and the equation ($E_c = 0$) is resolved semi-implicitly at the same time as the two momentum equations, thus replacing the boundary condition previously used in A_{out}.

For subcooled flow, the conditions listed in the above paragraph are used, but to avoid the numerical snags connected with rapid drop in critical velocity (e.g. oscillations and small time step), the model is extended in place of the two-phase model up to void fractions of 0.05. Finally, a transition for void fractions between 0.01 and 0.1 is used to move into the two-phase model. The only physical proof for this entirely numerical calculation is the data produced.

Continuing the discussion of Section 3, the model used for calculating critical conditions (thermodynamic equilibrium model) differs from the flow model (thermodynamic non-equilibrium model) used for fluid changes up to the critical section. In the numerical application of this critical condition, both the latter and its field of application are markedly altered.

5.2.3 TRAC CODE

The model used in the TRAC P1A and PD2 versions will merely be described.

5.2.3.1 Physical Model (TRAC PIA user's manual, 1979)

The critical flow model in TRAC /8/ forms part of the flow model. In the proximity of the break in the versions under consideration, the latter is a one-dimensional transient model with five equations:

- 2 equations of phase continuity,
- 1 momentum equation for the mixture,
- 2 equations of phase energy.

To this system of equations for calculating pressure, void fraction, average velocity of the mixture and different temperatures for the two phases, is added a "drift" equation for determining different velocities for each of the phases.

The system is in the following mathematical form:

$$AX'_t + BX'_z = C$$

Thus, flow is critical when one of the characteristic directions in the system is zero and the others positive, i.e. when det $|B| = 0$. Under this condition, the transfer terms are involved only if they include derivatives, which is not the case in the TRAC versions considered. Flow properties in the critical section are given by the fluid changes calculated by means of the previous flow model.

5.2.3.2 Numerical Calculation in TRAC

For one-dimensional components, TRAC uses a discretisation method with a set of meshes for scalar values and another offset one for vector values (see Figure 5.2).

TRAC assumes that the analytical form of the system of equations describing flow contains the critical condition that may be reached when the fluid changes are being calculated. Therefore, TRAC does not include a specific indication that critical flow occurs. Critical behaviour arises from the numerical properties of the discretised system.

At the break, however, the numerical treatment is different from the standard treatment:

- A discretisation mesh centred for the momentum equation is used whereas in the other components, the system is off-centre (for the reasons why this change is made, see Section 5.3.2).
- There is a shift from a semi-implicit to an implicit system for two reasons: longer time step, and stability with centred discretisation.

5.2.4 DRUFAN CODE

5.2.4.1 Physical Model

The critical flow model in the DRUFAN code /9/ is derived from the flow model, which assumes that the fluid is homogeneous (slip equal to 1). Steam is at saturation value and the liquid can be in thermodynamic non-equilibrium. The basic equations are thus:

- 2 equations of phase continuity,
- 1 momentum equation,
- 1 energy equation.

These equations are integrated over volumes for the entire circuit except at the break. At the break, a one-dimensional system is used with the same basic equations. This leads to the following system of differential equations:

$$AX'_t + BX'_z = C$$

The critical condition is obtained by applying the theory of characteristics, that is, through the condition $\lambda_i = 0$ with the other roots of the equation being positive, or equivalently when det $|B| = 0$.

Flow changes upstream from the critical section are calculated with the previous one-dimensional representation but under an almost steady state assumption, the system of equations is then merely:

$$BX'_z = C$$

The above arguments give the set of parameters that satisfy the critical condition. These parameters, therefore, depend on the mechanisms governing changes in flow, which in this case is primarily mass transfer between phases.

5.2.4.2 Numerical Calculation in DRUFAN

The one-dimensional calculation is done by means of a trial-and-error method. Pressure, enthalpy and the void fraction being specified, the inlet flow is adjusted so that the solution of the system satisfies exit critical conditions.

Coupling the one-dimensional calculation with that for the remainder of the circuit (averaged equations for a number of volumes) is straightforward since the basis for the equation systems is identical. The equations averaged for the volumes are solved. The mass flow found at the inlet into the one-dimension part is compared to the critical flow that would be given by the one-dimensional calculation with the same inlet conditions (flow determined by interpolation in previoiusly calculated data). If the mass flow found is lower than critical flow, the calculation proceeds further as before. If it is higher, critical flow becomes a limit condition instead of pressure. Flow is then systematically calculated by means of the one-dimensional model.

The one-dimensional calculation is included in the final volume. Thus, when entering the critical regime, the calculation with the basic model of the circuit may contain uncertainties in the final volume. The dimensions are too large in view of the considerable gradients that appear at this stage (this is precisely why a fine-mesh one-dimensional calculation is used). Owing to these uncertainties, the comparison between basic model flow and in the one-dimensional model flow may raise a problem in defining the transition to a critical regime.

5.2.5 CATHARE CODE

5.2.5.1 Physical Model

The critical flow model in the CATHARE code is incorporated in the flow model, which is a one-dimension transient model consisting of six equations (two-fluid model):

- 2 equations of phase continuity,
- 2 equations of phase momentum,
- 2 equations of phase energy.

This system calculates pressure, void fraction, and velocity and temperature for each of the two phases. The equation set has the following standard mathematical form:

$$AX'_t + BX'_z = C$$

Consequently, there is critical flow when one of the characteristic directions of the system is zero and the others positive, i.e. when det $|B| = 0$. This condition involves the derivative terms of the transfer laws. CATHARE uses the following as derivative terms:

- A virtual mass term in momentum transfers between phases.
- Terms ($\rho i \frac{\partial \alpha}{\partial z}$,$\rho i \frac{\partial \alpha}{\partial t}$) expressing the effects of pressure gradients between phases (these terms have little influence on the flow value but they are essential for describing laminar flow).

Flow properties in the critical section are given by the changes in the fluid as calculated by the flow model. Flow properties thus chiefly depend on mass and energy transfers and inter-phase drag.

5.2.5.2 Numerical Calculation in CATHARE

As with TRAC, CATHARE uses an offset mesh discretisation, but with two options, one semi-implicit and the other totally implicit. Two methods for numerically analysing critical flow are currently being tested:

- No special condition is laid down and critical flow behaviour results from the numerical behaviour of the discretised system. In this case, the calculation is semi-implicit or implicit with an off-centre mesh.
- The condition $\lambda_i = 0$ is used instead of the boundary condition on outlet pressure whenever it is found at the outlet junction during the calculation. In this case, the calculation is implicit with a system centred near the break.

We shall see in Section 5.3.2 that the second method is undoubtedly more accurate. Nevertheless, the first one is easier to use. The purpose of the current tests is to assess the errors introduced by the first method.

5.3 DISCUSSION

5.3.1 DISCUSSION AND COMPARISON OF PHYSICAL MODELS

In comparing physical models, the most frequently used classification divides them into those meeting the thermal equilibrium assumption or not, and those meeting the mechanical equilibrium assumption or not (slip between phases). This comparison can in fact be applied to many other thermohydraulic phenomena. Consequently, because of its general application, it is not suitable for pinpointing the specific physical features of critical flow with any accuracy.

In order to identify the mechanisms governing critical flow it is necessary to refer back to the definition of this process (see Ref. 6). A flow with fixed parameters at the pipe inlet, apart from flow rate, is critical if reducing the outlet pressure:

a) does not alter mass flow. The latter is then at a maximum;

b) does not alter any of the flow parameters upstream from a given section, referred to as critical section.

On the basis of this definition it can be demonstrated that critical flow appears when pressure changes cannot travel back up the flow, in other words when the velocities of these changes are either zero or positive in the critical section. From these basic properties arise two aspects which we shall describe more fully below:

- The local aspect within the critical section, where the choke actually occurs: flow thus meets a so-called critical condition.

- The evolutive aspect of the flow upstream from the critical section, these changes serving to produce the critical condition.

On the basis of these properties, two main types of models can be distinguished:

- Pre-integration models which assume a situation of no change leading to critical conditions.

- Models where the fluid evolution is calculated and there is a local critical condition.

5.3.1.1 Pre-integration Models

This first category includes RELAP4 models and many others described in the literature (by Fauske, Cruver, Zaloudek, and so on).

Evolution

These models are based on a "frozen evolution" immediately upstream from the critical section: e.g. isentropic evolution (HEM); the same liquid-steam ratio (HENRY); and isentropic evolution with slip (MOODY).

There is no physical justification for these assumptions. The more advanced models do not suggest that they be considered even as approximations. Furthermore, these assumptions are very often in conflict with the flow models used, especially since the criticality test is usually carried out by comparing outlet velocities calculated by the flow model and the critical flow model respectively. In other words, the two evolutions are based on different assumptions.

Critical Condition

The critical condition or conditions are written by analogy with the conditions fulfilled in single-phase and this can sometimes even be done directly in specific cases. This approach is justified for the sonic model or the HEM, which uses a flow model transposed from single-phase. Nonetheless, there are well-known limitations on the validity of this model, heat and mechanical imbalances forming a major factor governing two-phase flow.

As for the models attempting to consider either mechanical (MOODY) or heat imbalances (HENRY), this method leads to many critical conditions, almost as many as the authors themselves: maximisation of a given quantity; translation of the condition $\frac{dG}{dp} = 0$ with partial differential calculations varying from one author to the next; etc. In some models, two critical conditions are used instead of one to compensate for the additional degree of freedom provided by taking an imbalance into account. The model thus obtained can be used to calculate all flow parameters in the critical section and to establish an unique relationship between critical flow and stagnation conditions.

Concerning the physics of the models, establishing critical conditions by analogy is not a strict method, hence the large number of models obtained. Justifications and assumptions provided are usually vague and not based on physical or experimental facts.

Stagnation Conditions

The applications and principles of pre-integration models lead to the concept of stagnation conditions. However, for the models to be easy to apply, it is convenient to use a single common starting point for the assumed fluid evolutions. The simplest is the conditions in the fluid when brought to rest. Furthermore, critical conditions in these models attempt to define all parameters in the critical section so that there is then a general link between stagnation and critical conditions which can be used to draw up critical flow tables, for example.

The next step in depressurisation tests involves considering conditions as stagnation and adjusting the critical flow models to these data.

It is always possible to define stagnation conditions. This cannot be physically questioned - but the physical significance of these conditions may vary. It is essential to use them strictly, which does not usually happen in two-phase critical flow modelling.

Certain standard practices for two-phase flow (e.g. investigating the effect of L/D) would produce results in single-phase flow which would be difficult to apply and even erroneous, as they would not reflect actual physical events (see Appendix 5.1). Since the irreversibilities causing the difficulties are undoubtedly at least as important in two-phase than in single-phase flow, incorrect results in single-phase flow would obviously be even less meaningful in two-phase situations. In other words, the way stagnation conditions are used in two-phase critical modelling is not appropriate and questionable from a physical standpoint.

As the experimental level, it is clearly illusory to attempt to classify data in the form of flow as a function of reservoir or inlet conditions (see also MOBY DICK experimental data with and without a grid, Ref. 6).

In modelling, any attempt to calculate flow directly as a function of reservoir conditions would confuse the effects of flow evolution and critical conditions. This could lead to highly erroneous conclusions such as those on the mechanism of the L/D effect on critical flow.

The description of two-phase flow in the literature on compressible fluid mechanics shows the great advantage of using stagnation conditions because either tables or analytical formulae can be drawn up. This is possible only because the relationship between stagnation and critical conditions is virtually unique in physical terms. Critical velocity in gases is invariably almost identical with isentropic sound velocity. This is because for critical flow of gas, there is hardly any dynamic interaction between the pipe wall and the fluid. If there were, sound velocity would alter (see explanation on isothermal sound velocity in Ref. 6) and the relationship between stagnation and critical conditions would not be unique owing to the effect of the pipe wall. In two-phase flow we shall see in Section 5.3.1.2 that one of the major features is the existence of dynamic interactions between the phases.

As a result, the relationship between stagnation and critical conditions is bound to be extremely complex. Thus, involving stagnation conditions becomes even less attractive and reveals all the hidden drawbacks in pre-integration models.

5.3.1.2 Models with Calculated Evolution and Local Critical Conditions

The physical basis for these models is the one-dimensional flow model for calculating fluid evolution up to the critical section. The "critical conditions" is derived from the flow model itself. To be able to compare these models satisfactorily, a number of findings of the work done on the physical properties of such flow should be mentioned. For the sake of simplicity, we shall merely list the contents of the findings. Appendix 5.2 contains a more detailed explanation of their physical significance.

For single or two-phase flow meeting a number of clearly identified assumptions, the system of equations for one-dimensional flow can be written as follows:

$$AX'_t + BX'_z = C$$

Mathematical reduction of the system shows that transient or steady flow is critical (as defined in Section 5.3.1) if det $|B| = 0$ is locally true in the critical section. That is, if one disturbance in this section is stationary ($\lambda_i = 0$) while all the others move in the direction of the flow.

Physical analysis of the role of transfer terms shows that in addition to their usual algebraic part, a differential part (first order derivative) must be considered. Using this form of transfer, their effect on phase coupling can be taken into account at two levels:

- Evolutive coupling - this is the usual type. It mainly consists of the algebraic terms, which are dominant when the gradients are low or the transient slow (e.g. steady flow). There are the terms that govern inter-phase exchange which, as is known, determine the changes in two-phase fluid.

- Dynamic coupling - this is calculated by means of the differential terms. These alone couple the phases at the level of the critical condition and propagation velocities (the algebraic terms are not involved in these two processes, which proceed as though they were zero). Fixed evolution models (e.g. drift or thermal equilibrium models) are merely limit cases, where dynamic coupling expressed by a suitable form of the derivative terms is sufficiently strong for certain flow parameters to be so closely bound that they satisfy an algebraic equation. In addition, the stronger the dynamic coupling, the lower the critical velocities.

Through evolutive coupling (and to a lesser extent dynamic coupling), transfers determine the changes in the fluid up to the critical section. The set of parameters obtained in the critical section is not unique here but depends on the changes, so that the various effects can be physically taken into account during the changes (pipe length, special features, etc.). Within the critical section, this set of parameters will meet the critical condition involving transfers through dynamic coupling.

Critical flow arises here from the properties of the flow. These are included in the flow model, the mathematical "translation" of which is the general system of equations described above.

By taking these various mechanisms into account, it is then a simple matter to compare the various models with calculated evolution and a local critical condition, and to classify them so that the way in which the physical events are modelled is clearly reflected. For the codes described in Section 5.2, taking the two-fluid model without derivative terms as a basis for the transfer terms, the following table is obtained:

TRANSFERS	Dynamic Coupling			Evolutive Coupling		
	Mass	Momentum	Energy	Mass	Momentum	Energy
Two-fluid model* without derivative terms	no	no	no	Γ law	Inter-facial drag law	Heat transfer law
CATHARE	no	Virtual mass term	no	Γ law	Inter-facial drag law	Heat transfer law
TRAC-PD2	no	Drift flux	no	Γ law	set evolution law	Heat transfer law
DRUFAN	no	Slip=1	Steam at saturation	Γ law	set evolution law	set evolution law
RELAP5	Equilibrium	Virtual mass term	1 phase at saturation	Non-coherent flow model with critical model		

* e.g. K-FIX Code.

5.3.2 DISCUSSION AND COMPARISON OF NUMERICAL APPLICATIONS

5.3.2.1 Pre-integration Models

At the purely numerical level, pre-integration models do not raise any special theorectical problems. The potential drawbacks are those shared by the systems where not all quantities are calculated at the same point (here, centre of the volume and junction). As one of the aims of the models is to avoid having to give a detailed description of flow in the neighbourhood of the critical section, the final volume may be relatively large. Because of the ensuing high physical gradients, the error due to the points defining quantities being out of step can be considerable. Although we do not know whether there is a real correlation, RELAP4 MOD6 has problems calculating critical flow at a junction where the flow section is equal to the outlet volume (e.g. a double-ended break). It is then recommended to increase the transitional section artificially by this volume so that the calculation can be completed. This practice is obviously numerically suspect, especially since its possible effect on the accuracy of the data is not known.

5.3.2.2 Models with Calculated Evolution and Local Critical Condition

The DRUFAN and RELAP5 codes form a separate case due to their special numerical methods.

DRUFAN uses a steady-state one-dimensional module coupled with standard calculation by volumes and junctions. Whilst this method saves calculation time, inaccuracies may occur for transition to a critical regime due to inadequate meshing in the standard calculation (see Section 5.2.4). This drawback may have limited implications, however, because the transition to a critical regime is usually very rapid. The use of the steady-state module for describing transients may also require caution in some situations.

In RELAP5, tests are performed based on the standard flow calculation to determine whether flow is critical, and if it is, the pressure boundary condition is replaced by the critical one. Not only is the latter physically non-coherent, but also it requires various numerical operations. The numerical solution thus does not correspond to the analytical solution of the model, which normally invalidates the numerical method.

The TRAC and CATHARE codes both use offset mesh discretisation. In TRAC, there is no special condition for expressing critical flow. In CATHARE, two methods are currently being tested, one where no special condition is taken into account and the other where the critical condition is used to replace the pressure boundary condition when there is critical flow. These differences deserve to be explained.

For the analytical system, critical flow naturally seems like a property of the system behaviour (the system becomes singular). The question is whether the system shows the same property once it is discretised. Latrobe has shown that with an off-centre space discretisation system, the calculation could proceed beyond the critical condition without any singularity or uncoupling between the upstream and downstream sections arising in the discretised system. With a centred discretisation system, however, when the critical condition is reached the discretised system is singular in the critical section, so that upstream-downstream uncoupling may occur.

In TRAC, the discretisation system of one-dimensional components is off-centre except, as pointed out in Section 5.2.3, in the neighbourhood of the break, where a centre system is used for the momentum equation. Consequently, the code behaviour, which does not require any special condition to express critical flow behaviour, is justified.

In CATHARE, the first method now being tested and not involving any specific critical condition uses an off-centre system. In other words, the boundary condition remaining unchanged, the critical condition is reached without the system becoming singular. Flow even becomes supercritical at the outlet junction. On the other hand, the second method used a centred system and in effect the discretised system becomes singular at the outlet. The pressure boundary condition is then replaced by the critical condition. Obviously, the second method is far more accurate but the purpose of the current test is to evaluate the errors made with various methods before selecting the one providing the best compromise between accuracy and efficieny. The smaller the meshes in the neighbourhood of the break, the smaller the differences between methods.

5.3.3 DISCUSSION OF THE MAIN TRENDS SHOWN BY THE RESULTS

The results of the models have been very broadly compared with experimental data. There are two types of comparisons:

- Those made with a view to verifying the codes. In principle such comparisons are done on a fixed model. Nevertheless, a distinction should be made between first generation codes such as RELAP4 (where options are available and when applying such options, the user may appreciably alter the results of the calculation) and advanced codes, such as TRAC, RELAP5, DRUFAN and CATHARE (where the models are non-adjustable).

- The model itself is adjusted, which generally produces good test-calculation agreement. Such comparisons are usually difficult to compare with one another. Furthermore, due to the wide range of models, it is not really possible to draw conclusions about good agreements with experimental data. This is why this chapter has been restricted to the models used in a few safety codes.

Owing to the differences of principle between pre-integration models and models with calculated evolution and local critical condition, we shall examine the results of both types of models separately as in the previous section.

5.3.3.1 Pre-integration Models

These are mainly RELAP4 models. When applying these models, users are free to choose the following:

- The multiplier C_D (sometimes referred to as contraction coefficient).

- The field of application of the models (transition point).

The only way to act directly on the models by changing the programming is with the HENRY-FAUSKE model, where the value of parameter N can be adjusted. Because of the principles on which they are based, the other models (HEM, sonic, MOODY) do not have any adjustable coefficients and are, therefore, fixed.

In view of the fields covered by the models and users' experience, the most frequently applied set of models is :

HENRY-FAUSKE	sub-cooled ($X_T < 0$)
HEM	for $X_T > 0.02$
Transition	for $0 < X_T < 0.02$

MOODY is sometimes used in two-phase studies for assessment calculations to meet licensing regulations, or in the steam phase as it produces results equivalent to those of HEM.

In comparison with experimental data, it is soon obvious, as noted by all RELAP4 users, that the above set of models is inadequate for reproducing tests. It is necessary to adjust the multiplier C_D. Thus, for a given test, it is sometimes possible to find the factor which will yield data fairly close to flow values just outside the break. Attempts have even been made by D.G. Hall (Refs 10 and 11) to determine optimum C_D coefficients for various experiments and various tests on a same experiment. Data indicate broad scattering of C_D coefficients, depending on geometry (difference between experiments) and flow conditions (difference between tests on the same experiment).

The discussion on physical models in Section 5.3.1.1 suggests that by adjusting the coefficient C_D, an attempt is made to correct modelling flaws. An example is work by E.J. Martinec /12/ done at the same time as that by D.G. Hall /11/. The latter sought C_D coefficients which would yield the MARVIKEN data, test by test. The former determined a coefficient N of the HENRY-FAUSKE model which would also yield the MARVIKEN data. Martinec concluded that N must be increased, compared to the value used by Hall, which the latter had compensated by C_D values not equal to 1.

The conclusions from these few examples and the general experience of such models are:

1) These models cannot be used without multipliers.

2) By optimising the multiplier, the correct flow values are obtained in a given specific situation, but not necessarily the other parameters such as critical pressure. This optimisation is by no means general.

3) For a transient calculation on a system, accuracy will also depend on the accuracy of the model's inlet conditions (stagnation conditions).

4) Fairly reliable predictions can be made by applying these models to tests simulating large breaks. For small break simulations, this varies enormously due to the added difficulty of identifying upstream conditions owing to phase separation phenomena.

5.3.3.2 Models with Calculated Evolution and Local Critical Conditions

For these models, all physical data are concentrated in inter-phase transfer laws, which operate as follows:

- The algebraic and differential part influence flow changes and thus flow parameters within the critical section.

- The differential part (or in a concealed way, the number of equations) influences the critical condition.

Much work has been done to establish suitable forms for these laws. For the algebraic parts, analytical (e.g. based on bubble growth data) or semi-analytical models (e.g. relaxation model adjusted to tests) have been developed. The differential terms that are usually introduced include virtual mass terms in the momentum equations and a $\frac{Dp}{Dt}$ term in the mass transfer. Obviously, "drift-flux" type or homogeneous models can be reduced to specific differential terms. Good test predictions can be made with these models for several parameters: flow, pressure in the critical section, and sometimes pressure and void fraction behaviour.

Nevertheless, coherence of the data is a problem as for any comparison between calculations and experimental geared towards adjusting certain model parameters. In this context, good results are obtained from very different physical assumptions. For critical velocities and the values of flow parameters within the critical section, the various models show the following:

K-FIX (two-fluid model without derivative terms)	Very high critical velocity. Flow parameters: $V_V \neq V_L$ $T_V \neq T_L \neq T_{SAT}$ (Considerable energy exchange and high interfacial drag).
CATHARE	Slightly higher critical velocity than the frozen model. Flow parameters: $V_V \neq V_L$
TRAC	Critical velocity $\neq$ frozen model. Flow parameters slip = drift-flux.
DRUFAN	Critical velocity = frozen model. Flow parameters slip = 1.
RELAP5	Critical velocity: $\neq$ homogeneous model (since the frozen model does not give good results according to the authors). Flow parameter $V_V = V_L$.

These examples, classified in decreasing critical velocity order, cover the entire range of potential velocities in the homogeneous version of the two-fluid model. An obvious parallel can be made with transfer laws giving very different flow parameters in the critical section with slip ranging, for example, from the virtually homogeneous to the drift-flux model. This leads to uncertainty regarding the respective weight of the algebraic and derivative terms.

The more sophisticated physics in these models generally leads to much more convincing test-calculation agreements than with pre-integration models. Nevertheless, these are "young" models and are, therefore, less reliable for flow predictions than pre-integration models, which have been much more extensively adjusted. The LOBI-PREX exercise is a good example since the best results were obtained with RELAP4. Nevertheless, calculated evolution models are much more promising for scaling up the data to reactor conditions, whereas RELAP4 models lack physical pertinence.

5.4 RECOMMENDATIONS

5.4.1 PRACTICAL CALCULATIONS: TEST AND SAFETY CALCULATIONS

Use of the various models for test calculations shows those in RELAP4 to be the most reliable. Where it is possible to adjust C_D to a separate test representing the break, the data obtained can be relatively good, except perhaps for certain test conditions (e.g. transition from large to small breaks).

For safety calculations, the biggest problem is that of scaling up the data. A dilemma arises here:

- whether to use RELAP4 type models, which are not very reliable for translating data and can only be used under limited conditions but are by now sufficiently well "tuned" to give fairly reliable experimental comparisons; or

- more physical models in advanced codes, which are reliable for translating data but less reliable in tests.

The first option may be preferable at the moment, pending development of better models for predicting the major consequences of critical flow from the safety standpoint.

One such consequence for large breaks is distribution flow either side of the break (in the case of breaks not clearly double-ended or "button-hole" breaks) where shifting the stagnation point may markedly alter cladding temperatures. For small breaks it is also important to know mass flow and enthalpy flow at the break. Mass flow will give the quantity of water remaining in the circuit whereas enthalpy flow will determine whether it is possible to balance the three energy sources and sinks in the circuit, namely the steam generators, the core, and the break.

In short, the following recommendations can be made:

1. *In order to obtain fairly reliable experimental predictions, it is probably preferable to use RELAP4 models in spite of their drawbacks.*

2. *In the absence of adjusted models with greater reliability for scaling up data, the RELAP4 type models are the only suitable ones for use. Sensitivity studies are necessary to mitigate any scaling errors.*

5.4.2 PHYSICAL MODELLING

To better understand physical mechanisms occurring in critical flow and thus acquire a general description that can be more safely scaled up, in our view it is essential to continue development work on calculated evolution models. This should cover both types of transfer terms:

- Algebraic, which are the dominant terms for low gradients and slow transients. More information on these transfer terms should be useful for many other phenomena that are still poorly understood in both large and small breaks.

- Differential, which determine critical velocity through dynamic coupling of phases. Their role is important and the couplings need to be understood more clearly. Coupling under the assumption of equal phase pressures is especially important, as mentioned in Appendix 5.2, and is a weak point in current models. Thus, it is important to continue work on this type of coupling, as started by Boure /13/ for example. It is also necessary to evaluate the relative importance of differential and algebraic transfer terms in the description of critical flow. Thus, it might be useful to compare the calculations of models with very different relative fractions of dynamic and evolutive couplings. This could be done as a standard problem by a group of model designers.

Further improvements to physical modelling based on calculated evolution and local critical condition should be provided by a better description of the specific features generally proceeding the break. Examples are the cross-section restrictions in the experimental representation of large breaks, or tapping points on the main pipes in the case of small breaks (pipes where laminar flow could occur). Physical modelling on these lines is the only possible basis for writing a 2D code for evaluating the importance of two-dimensional effects which might occur in real situations, e.g. not simple breaks and "button-hole" breaks.

In a second stage, it might be useful for numerical efficiency purposes to return to pre-integration models, but these should be written on the basis of calculated evolution models as approximations of the latter.

In short:

1. *It seems essential to continue development work on calculated evolution models.*

2. *This research should cover both algebraic and differential transfer terms. A calculation comparing the various models could be handled as a standard problem, focusing on the relative importance of the two types of terms.*

3. *Better evaluation of coupling through pressures appears to be important and would improve one of the weak spots in current models.*

4. *Special attention should be paid to the calculation of the specific features that are generally found upstream of the break.*

5. *If it turns out to be necessary to develop pre-integration models for numerical efficiency reasons, they should be established as an approximation of calculated evolution models.*

5.4.3 NUMERICAL APPLICATION

As indicated, two numerical problems must be overcome in calculating critical flow:

- The gradients of the variables are very high, so that mesh size must be adequate and sensitivity studies must consider this.

- The discretised system may behave differently than the analytical one in the neighbourhood of the critical section. If so, it may be necessary to carry out ad hoc tests to ensure that the correct numerical interpretation of flow criticality has been made.

In summary, we shall repeat the general recommendation, which is still far from universally observed:

The numerical method must yield the solution to the analytical system or a controlled approximation of the solution. Special attention must be paid to accuracy problems due to the high gradients and the criticality problem.

5.4.4 CONFIRMATORY TESTING OF MODELS

To ensure that a model does not correctly predict a parameter (e.g. mass flow) by making several countervailing errors, it is essential to check the model on all parameters measured throughout the test. This will indicate whether a model is truly physical and if it can, therefore, be validly used for scaling up to reactor size (it should be possible to scale up a number of countervailing errors!). Confirmatory testing of pre-integration models solely based on mass flow is, therefore, not adequate. A good prediction of critical pressure is a minimum requirement. For calculated evolution models, tests should involve any measured parameter that forms a control point for the model, which usually calculates more parameters than can be measured. Confirmatory testing of critical flow models must not be restricted to one single parameter (especially mass flow) but must be based on all parameters measured.

5.5 CONCLUSIONS

In spite of all the theoretical and practical research on critical flow, there are no models that can reliably predict this phenomenon. One of the reasons is that the work done is often insufficiently exact. On the basis of one-phase flow data, which were not always fully understood, many theories have proceeded by analogy without questioning whether the analogy used was well-founded. There are as many analogies, and hence models, as authors. In these models, correlation work has been done either directly or through the multiplier. This approach is often regarded as empirical or "engineering" and is constrasted with another, physical-mathematical and more theoretical, approach.

Such a distinction does not make sense: any fluid mechanics theory and, especially, any two-phase flow theory ends with an empirical correlation. The difference between the studies contrasted here concerns the level at which the correlation occurs, e.g. either coefficient C_D or the number of bubbles and their initial diameter in a mass transfer law. Nothing should, therefore, separate the scientific approaches adopted in both cases. It is vital for them to be followed with the same rigour. Paradoxically, the so-called empirical approach is the one that possibly requires more rigour because the true physical factors determining whether the phenomenon can be correlated must be fully identified. In the critical flow analysis we have just made, the more empirical approach, which involves pre-integration models, has not really identified the true physical mechanisms and this is why the models cannot be fully correlated.

This undoubtedly constitutes grounds for using calculated evolution models in advanced codes. The need for further studies on these models arises since it is a much more physical description. Instead of correlating effects, an attempt is made to correlate the causes of the effects, i.e. the transfer laws, in the hope for data that can be more reliably scaled up. The work on transfer laws thus contributes to the general knowledge about two-phase flow. Critical flow becomes just a specific property of flow about which it is very important to acquire physical information. The least-understood transient situations include injection phases in small and large breaks, reflooding in large breaks, stratification and condensing reflux operation in small breaks. In all these situations the role of two-phase flow is crucial.

Critical flow provides excellent tests for models because of its complexity. Apart from the need for better prediction of critical flow, this forms an added incentive for further studies on the subject.

LIST OF SYMBOLS USED

Note

The same symbol may be used for more than one magnitude, but there is no ambiguity if the definition used is related to the context.

CAPITAL LETTERS

A	Cross-section area
A	Coefficient matrix
B	Coefficient matrix
C	Right-hand side vector
C_D	Flow multiplier or contraction coefficient
D	Pipe diameter
G	Uniform mass flow
J	Junction
L	Length
M	Proportionality constant in the HENRY-FAUSKE model
T	Temperature
X	Flow parameter vector
V	Velocity
$\mathcal{V}$	Volume
V	Centre of volume

SMALL LETTERS

a	Sound velocity
p	Pressure
t	Time
z	Space co-ordinate

GREEK LETTERS

α	Void fraction
γ	Slip
Γ	Mass transfer
λ	Characteristic velocity
ρ	Specific gravity

SUBSCRIPTS

S	Isentropic
C	Critical
u	Interfacial
i	Numerical index
j	Numerical index
L	Liquid
m	Mixture
sat	At saturation
t	Temporal
T	Thermodynamic
V	Steam
z	Spatial

UPERATORS

'	derivatives
det | |	determinant

Figure 5-1

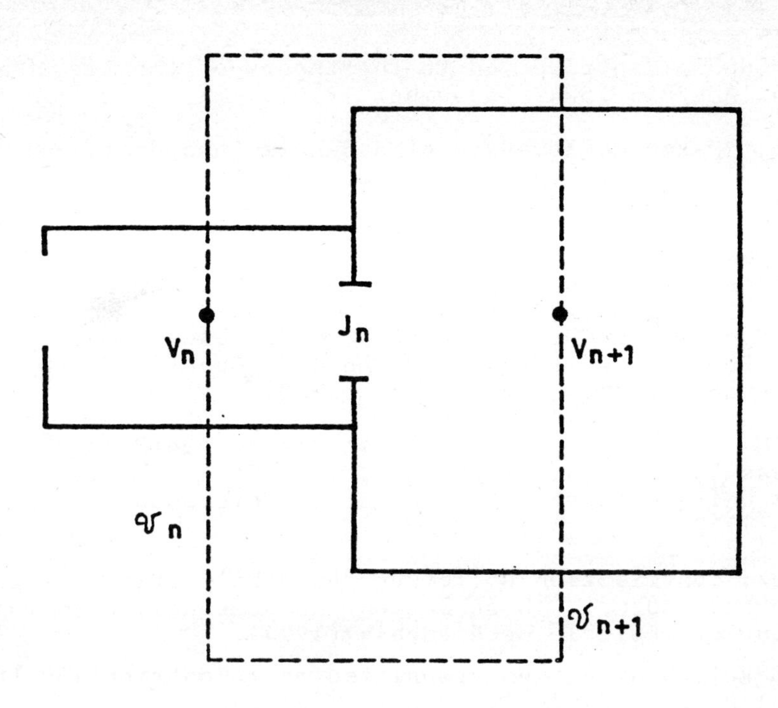

Figure 5-2

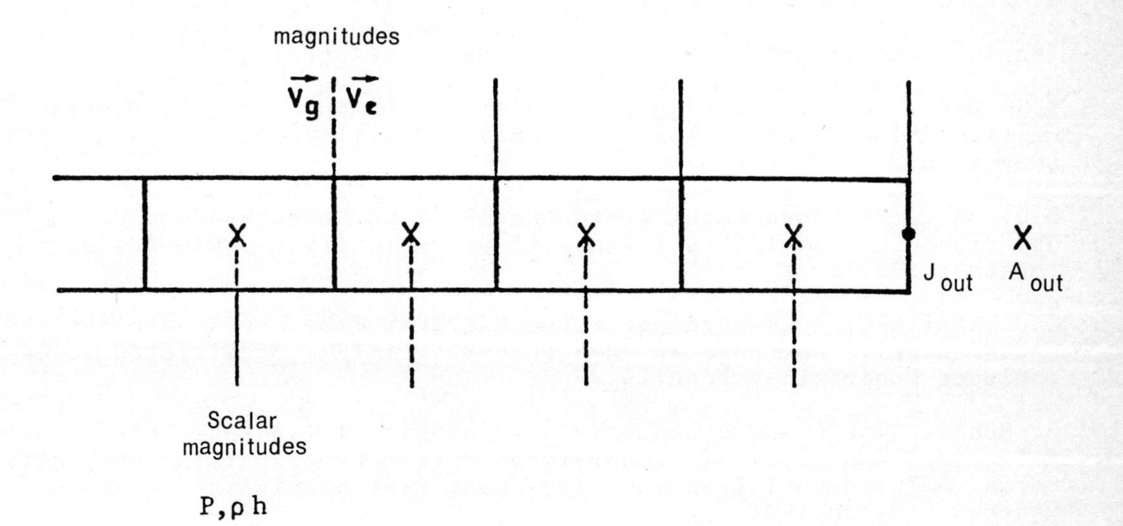

REFERENCES

/1/ M.N. HUTCHERSON, Contribution to the theory of the two-phase blowdown phenomenon. AML/RAS 75-42 (November 1975).

/2/ F.J. MOODY, "Maximum Flow-Rate of a Single Component, Two-Phase Mixture". Journal of Heat Transfer, Trans ASME 87 (February 1965).

/3/ R.E. HENRY, H.K. FAUSKE, "The Two-Phase Critical Flow of one Component Mixtures in Nozzles, Orifices and Short Tubes". Journal of Heat Transfer, Trans ASME 93 (March 1971).

/4/ V.H. RANSOM, J.A. TRAPP, "The RELAP5 Choked Flow Model and Application to a Large Scale Flow Test". ASME Nuclear Reactor Thermal-Hydraulic Topical Meeting, Saratoga, New York (1980).

/5/ J. BOURE, M. REOCREUX, "General Equations of Two-Phase Flows. Applications to Critical Flows and to Non-Steady Flows". From Fourth All Union Heat and Mass Transfer Conference, Mixisk, May 15-20 (1972).

/6/ M. REOCREUX, Contribution à l'étude des débits critiques en écoulement diphasique eau-vapeur. Thèse Docteur-es Sciences, Université scientifique et médicale de Grenoble (1974).

Vol. 2 and Vol. 3 have been translated as : Contribution to the Study of Critical Flow Rates in Two-Phase Water Vapor Flow. NUREG-TR-0002 (Vol. 2) (1977), NUREG TM-0002 (Vol. 3) (1978).

/7/ V.H. RANSOM et al, RELAP5/MOD1 Code Manual. NUREG/CR-1826 - EGG-2070 (March 1981).

RELAP4/MOD5. A Computer Program for Transient Thermal-Hydraulic Analysis of Nuclear Reactors and Related Systems - User's Manual. AMCR-NUREG 1335 (September 1976).

/8/ TRAC-P1A. An Advanced Best Estimate Computer Program for PWR LOCA Analysis. NUREG/CR-0665 - LA 7777-MS (March 1979).

/9/ K. WOLFERT, "A New Method to Evaluate Critical Discharge Rates in Blowdown Codes that are based on the Lumped-Parameter Technique". Thermal Reactor Safety Meeting, Sun Valley (August 1977).

/10/ D.G. HALL, "Empirically Based Modelling Techniques for Predicting Critical Flow Rates in Nozzles, Tubes and Orifices". CVAP-TR-78-010 (May 1978).

/11/ D.G. HALL, "An Assessment of the RELAP4/MOD6 Computer Code Using Data from the MARVIKEN CFT Project" (Proprietary), EGG-GAAP-5032 (October 1979).

/12/ E.J. MARTINEC, A Comparison of the MARVIKEN flow tests with the HENRY-FAUSKE model. AML/RAS/LWR 74-8 (December 1979). /Restricted release: includes Proprietary results./

/13/ J. BOURE, "The Closure Laws of One-Dimensional Two-Phase Flow Models: A Critical Appraisal of Algebraic and First-Order Partial Derivative Forms". Third CSNI Transient Two-Phase Flow Specialists Meeting, Pasadena (March 1981).

/14/ J.M. DELHAYE, "Contribution à l'étude des écoulements diphasiques eau-air et eau-vapeur". Thèse Docteur-es Sciences, Faculté des sciences, Université de Grenoble (1970).

APPENDIX 5.1

EXAMPLE OF THE USE OF STAGNATION CONDITIONS IN SINGLE-PHASE FLOW

Consider a reservoir filled with gas under pressure flowing out through a constant cross-section pipe without heat exchange with the wall but with drag and, for simplicity's sake, under steady conditions. It is known that throughout the pipe the stagnation conditions for the gas vary owing to the variation in entropy due to drag. Under steady conditions, flow is constant and if there is critical flow at the outlet, this will be due to choking (see Ref. 6), which can be correlated to variations in the stagnation conditions of the gas. Since there is a relationship between stagnation and critical conditions of gases, which is a good approximation of the real facts, the stagnation conditions corresponding to critical flow at the outlet can be determined. Since stagnation conditions at the outlet are different to those inside the reservoir, the critical relationship does not apply between reservoir conditions and critical ones. Furthermore, if while reservoir conditions remain unchanged pipe length is increased, the flow leading to critical conditions at the outlet will be lower and stagnation conditions at the outlet will also be different. Consequently, if for gases critical flow modelling had been carried out in relation to reservoir conditions, as some researchers are doing or wish to do with two-phase flow, the conclusions would have been an L/D effect. This example clearly shows that any attempt to incorporate this effect directly into a relationship between critical flow and reservoir conditions is bound to lead to a mere approximation which would not reflect physical mechanisms.

It also shows that the use of stagnation conditions in single-phase flow is a procedure fully understood by its users. The above-developed extension of a two-phase use to a single-phase case clearly shows the amount of confusion in some of the work on two-phase critical flow modelling as regards the concept of stagnation conditions.

APPENDIX 5.2

SUMMARY OF THE PHYSICAL ANALYSIS OF MECHANISMS GOVERNING SINGLE- AND TWO-PHASE CRITICAL FLOW

This summary is chiefly based on the analysis given in volume 1 of Ref. 6.

1. ESTABLISHMENT OF SUITABLE CONDITIONS FOR CRITICAL FLOW

By means of several clearly identified and discussed hypotheses /5, 6, 14/, the system of equations describing transient one-dimensional two-phase flow can be written as follows:

$$A\,X'_t + B\,X'_z = C$$

where X is the column vector for flow variables, A and B matrices and C the column vector for the transfer terms.

For any flow described by a system of this type (which also includes single-phase flow), it has been shown under steady flow /6/ that for the flow to be critical it is necessary that within a section, the relationship det $|B| = 0$ should hold. Simultaneously, a compatibility condition $N_i = 0$ involving the transfer terms and determining the location of the critical section must also hold.

If disturbances are propagated, the theory of characteristics shows that propagation velocities are obtained by resolving the equation det $|B| = 0$. There is critical flow if the disturbances cannot move back against the flow upstream from the critical section, in other words if within this section solutions λ_j of the characteristics equation are zero for one (λ_i) and positive for all the others. Clearly then if $\lambda_i = 0$ then det $|B| = 0$. The condition det $|B| = 0$ established under steady conditions, therefore, also applies under transient conditions.

At this point in the summary of the analytical results, it should be noted that the critical condition obtained applies both to steady and transient conditions and both to single- and two-phase flow, and that it is not superimposed on the flow model but a property of the model itself. This represents a considerable advance over the critical conditions previously used in pre-integration models, where such conditions often bore little relationship with the assumed evolution of the fluid upstream from the critical section.

2. ROLE OF TRANSFERS BETWEEN PHASES AND BETWEEN PHASES AND THE PIPE-WALL

Continuing the analysis /6/ from the described results above clearly identifies the role of the transfer terms.

Until recently, it was usually assumed that the transfer terms did not contain derivative terms. In other words, all the transfers are grouped in vector C. Propagation velocities and the critical condition are thus in no way dependent on transfer terms since the mathematical relations governing them involve only matrices A and B. Consequently, for these phenomena it is as if these terms were zero locally, in other words as if the phases were dynamically uncoupled at the transfer level. The following example will give a clearer idea of this dynamic uncoupling.

Let us consider a flow model using two continuity equations whose right-hand side (C) contains mass transfer Γ. If its form is purely algebraic, Γ does not intervene at the critical condition level which would thus be the same as if $\Gamma = 0$. The critical flow obtained then corresponds to that of the so-called "frozen" model, i.e. without mass transfer. Dynamically this means that at the critical point there is no inter-phase interaction through mass transfer, notwithstanding the fact that during flow evolution mass transfer and hence coupling occurs between phases owing to the algebraic portion of Γ.

This dynamic inter-phase uncoupling at the critical point and propagation velocities may on occasion produce unrealistic physical data. It, therefore, seems necessary /5/ to see how dynamic coupling could be incorporated. By examining several models and their physical significance, the author /6/ was led to the concept of derivative terms within transfer terms.

Owing to these derivatives, the transfers modify matrices A and B and hence the expressions for propagation velocities and the critical condition. At the point of the critical condition or any propagation of disturbances, the two phases would thus no longer behave independently as though transfer were zero but as though there were instantaneous (or dynamic) transfer corresponding to the differential portion. We then have evolutive coupling through the algebraic portion of the transfer terms acting only at flow evolution level and dynamic coupling through the differential portion acting also at evolution level but being the only one to influence the critical condition and dynamics of the disturbances.

Depending on the derivatives introduced, dynamic coupling will vary in strength. It reaches a maximum when it leads to "static" evolution whereby an algebraic evolution law applies between the fluid parameters. The system of equations describing flow then degenerates into a system containing one equation less but to which the evolution law must be added (see Ref. 6). By introducing derivative terms it is thus possible to understand the physics of models containing a smaller number of equations to which evolution laws are added (e.g. slip equal to 1, thermal equilibrium and so on).

Taking once more the above example of the model with two continuity equations, if mass transfer Γ contains derivative terms, this will introduce dynamic coupling through mass transfer at the point of the critical condition, which will thus differ from the condition in the frozen model (as though $\Gamma = 0$)obtained with Γ in algebraic form. Obviously this dynamic coupling will make the phases tend towards thermal equilibrium. The limit case will be reached when the derivatives are such that the flow always remains in thermal equilibrium. The two physical continuity equations can then be reduced to the continuity equation of the mixture plus an algebraic evolution law ($T_V = T_L$). This example thus shows how dynamic coupling through mass transfer can be a way of passing from the frozen to the equilibrium model.

At this stage in the analysis, an assumption made in all present models and often overlooked should be noted. It concerns equal phase pressures. Assuming that, at any moment in time, phase pressures are equal irrespective of the disruption affecting the fluid, it can be readily presupposed that there is very strong dynamic coupling between phases /6/. This is probably a weak spot in present models (see also Ref. 13).

3. IMPLICATIONS FOR THE DESCRIPTION OF CRITICAL FLOW IN CURRENT MODELS

If the various models in current usage are now examined, the one involving the weakest dynamic coupling is the two-fluid model without derivatives for transfers. The critical condition is then the same as though there was no mass, momentum or energy transfer.

From this description with very little coupling, an initial dynamic coupling can be introduced at the level of the momentum transfer. This is what happens when a virtual mass term is considered (CATHARE, RELAP5 This coupling tends to connect velocities in the two phases. If it becomes stronger either an algebraic law between the two velocities is being applied (e.g. drift flux model) or the velocities becomes equal (homogeneous model). In the last two cases, the two momentum equations are reduced to a single equation plus a relation (drift or equal velocities).

A second dynamic coupling can be introduced at the mass transfer level so that in the previoius example, it is possible to pass from the frozen to the equilibrium model.

A study of the effect of dynamic couplings on critical velocity shows (see Figure 5.3) that the stronger the coupling, the lower the velocities /6/. For the two-fluid model without derivative terms, velocities are very high (ranging between sound velocity in steam and sound velocity in liquid) which is unrealistic. Dynamic coupling through momentum transfer leads to lower velocities down to the frozen model values. If dynamic coupling through mass transfer is then added, critical velocities are reduced still further to those of the homogeneous equilibrium model, which are the lowest of all models and correspond to maximum dynamic coupling of phases ($T_V = T_L = T_{sat}$, $V_V = V_L$).

Having summarised these results, it is now necessary to examine how fluid evolution and critical conditions interact. For a system with n equations, the condition det $|B| = 0$ gives a relation for the flow variable n. From this, flow can be deduced as a function of the (n-1) other variables. The values of the latter will be determined by resolving the system $A X'_t + B X'_z = C$ and will thus strongly depend on the algebraic and differential components of the transfer terms.

The various mechanisms determining flow evolution and the interaction between the evolution and the critical condition can be represented as follows:

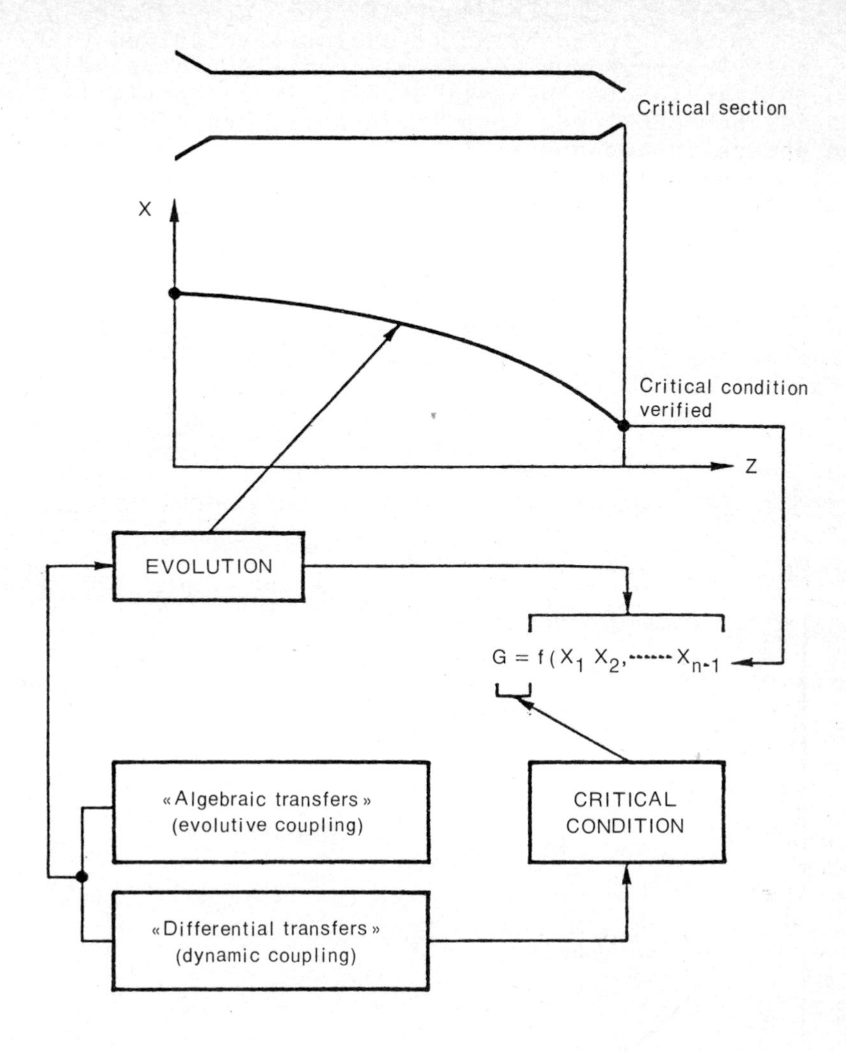

All the transfers determine fluid evolution, which will give the set of values X_1, X_2 ... X_{n-1} whereas transfers in differential form will alone influence the critical condition and the form of function f.

As can be seen, unlike pre-integration models, there is not just one set of flow variables which can lead to a given critical flow. The set of variables depends on flow evolution, so that it is possible to take the effects of pipe length and various specific features of the flow (obstacles, narrowing, inlet effects) physically into account.

Note

The above analysis also applies to gaseous single-phase flow. In that case, vector C only contains terms for transfer to the wall. These are usually expressed algebraically, which is why critical velocity is always equal to the isentropic sound velocity even if the evolution of the fluid is not isentropic owing to evolutive coupling with the wall. This result usually corresponds to actual conditions because if dynamic coupling through transfer to the wall occurs, it is generally weak and it might very easily be much lower than the dynamic coupling that might exist between phases in two-phase flow. In order to obtain a critical velocity equal to the polytropic or isothermal sound velocity, such couplings must occur (see Ref. 6).

Figure 5-3 **RATE OF CRITICAL VELOCITIES FOR VARIOUS MODELS**

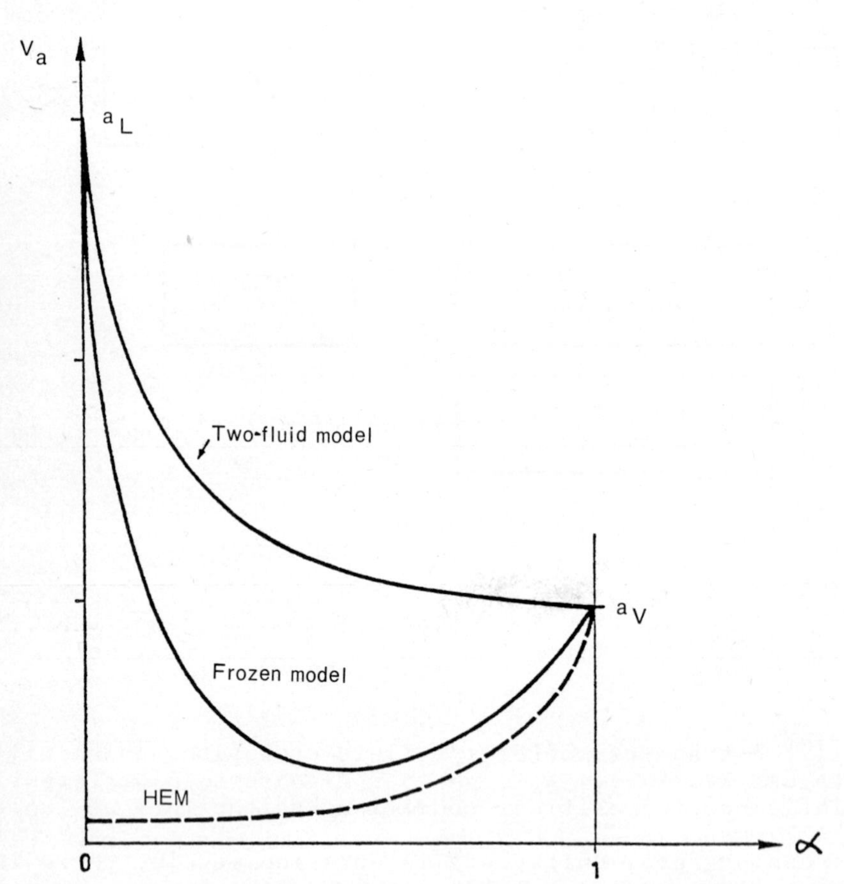

6

GENERAL CONCLUSIONS AND RECOMMENDATIONS

Each of the three preceding chapters includes conclusions and recommendations reflecting the views of the individual authors. In this chapter we attempt to draw overall conclusions and make recommendations on the most profitable lines of experimental and theoretical research. There is some conflict between the conclusions of the earlier chapters. These conflicts have deliberately been allowed to remain in order to emphasise the incomplete nature of the state-of-the-art, and thus the need for further work. The conclusions in this chapter, however, represent a consensus view of all the experts involved in the production of this report.

6.1 CONCLUSIONS

The requirement to predict critical flow in single and two-phase water has been identified for three main areas. Firstly, it is necessary to calculate reactor blowdown under LOCA conditions, not only in an average sense but also with a correct time (or enthalpy) variation. The absolute accuracy requirement of this application is not great because the break size is indeterminate. Secondly, the planning and analysis of experiments require critical flow calculations, in particular in 'Standard Problem' type exercises which are blind. An alternative is to use measured flow as boundary conditions in the calculations, but the measurement of two-phase flows under critical conditions is difficult and subject to inaccuracy. Thirdly, critical flow is an essential component in the calculation of forces on structures. At present, very conservative methods are used, and there may be an economic benefit in refining methods of calculation.

A large number of critical flow models exist, ranging from empirical correlations through physically-based models of idealised situations to full two-fluid models. A major problem in selecting the best models is that each originator has been able to find data which is well-fitted by his own model, but extension to other data is often disappointing.

The empirical/idealised models are sometimes thought of as an 'engineering' approach, in contrast to the physical/mathematical approach of the two-fluid models. The models in the first category are often based on the concept of fluid stagnation conditions, which makes them difficult to apply in many practical situations.

Two-fluid models have a fundamental appeal in that they use basic microscopic physical concepts, and, therefore, should be relatively insenstivie to the specific geometrical situation. However, it is important to realise that these models also rely ultimately on empirical correlated data, and in this sense are no different from the first category.

At present, the state-of-the-art is such that the requirements set out above are best met by properly correlated models of the so-called engineering approach, for instance the models in RELAP4. However, to get the best results from these models it is necessary to make final adjustments by using a discharge coefficient C_D different from unity.

A large amount of experimental critical flow data exists. Most of the data are for pipes, nozzles and orifices of less than about 15 mm diameter. The only data for diameters of >200 mm are those from MARVIKEN, which range up to 500 mm diameter. Little data exist for other components such as valves, or for piping failures other than guillotine breaks.

The reported experimental parameters are often incomplete and inconsistent. Several authors did not record sufficient information to allow the fluid stagnation conditions to be specified.

The logical extension of this work would be to make a detailed comparison of all the models and data found. However, this represents much too large a task for this SOAR, and would be of doubtful value in the light of the quality of much of the data.

6.2 RECOMMENDATIONS

Empirical "engineering" models of the type found in RELAP4 are currently recommended for both the planning and analysis of experiments and for plant safety calculations. However, for best results values of the multiplier C_D different from unity will be required. For safety calculations the uncertainty in scaling up to the full-size plant must be covered by sensitivity studies, though this is necessary in any case to cover uncertainty in the geometry of actual breaks.

Future theoretical effort should be concentrated on the mathematical two-fluid model approach, because only this approach offers the prospect of methods which will be insensitive to geometry and, therefore, extrapolate reliably to full-size plant. Current advanced codes, e.g. TRAC and RELAP5, attempt to use the two-fluid model approach, but they are incomplete or contain arbitrary terms. A satisfactory model will not be achieved without the inclusion of differential interphase transfer terms. It is also a desirable feature that the equation set being used at the critical flow point is identical to that used elsewhere in the circuit. If, for computational efficiency, some form of pre-integration model becomes necessary then this should be an approximation to the full model with identifiable errors.

Future experimental effort should concentrate on providing data which will aid the development of the two-fluid models and provide the necessary constitutive relationships. This will require a change in the approach shown in the majority of experimental reports to date in that it will be necessary to measure as many parameters as possible, and not just the overall mass flow rate. Close liaison between the experimentalists and the model developers will be necessary to ensure that the right parameters are measured and reported. It should be noted that such information will be directly relevant to two-fluid modelling in general, and not just to critical flow calculations.

The model developments envisaged here are unlikely to produce routine analysis and assessment methods in the short terms, and there may well be a need for further ad hoc critical flow measurements, especially on components such as valves and pipe breaks of the slit type.

SOME NEW PUBLICATIONS OF NEA

QUELQUES NOUVELLES PUBLICATIONS DE L'AEN

ACTIVITY REPORTS

Activity Reports of the OECD Nuclear Energy Agency (NEA)

- 9th Activity Report (1980)
- 10th Activity Report (1981)

Free on request – Gratuits sur demande

Annual Reports of the OECD HALDEN Reactor Project

- 19th Annual Report (1978)
- 20th Annual Report (1979)

Free on request – Gratuits sur demande

RAPPORTS D'ACTIVITÉ

Rapports d'activité de l'Agence de l'OCDE pour l'Énergie Nucléaire (AEN)

- 9e Rapport d'Activité (1980)
- 10e Rapport d'Activité (1981)

Rapports annuels du Projet OCDE de réacteur de HALDEN

- 19e Rapport annuel (1978)
- 20e Rapport annuel (1979)

• • •

INFORMATION BROCHURES

- OECD Nuclear Energy Agency: Functions and Main Activities
- NEA at a Glance
- International Co-operation for Safe Nuclear Power
- The NEA Data Bank

BROCHURES D'INFORMATION

- Agence de l'OCDE pour l'Énergie Nucléaire : Rôle et principales activités
- Coup d'œil sur l'AEN
- Une coopération internationale pour une énergie nucléaire sûre
- La Banque de Données de l'AEN

Free on request – Gratuits sur demande

• • •

NUCLEAR ENERGY PROSPECTS

Nuclear Energy Prospects to 2000
(A joint Report by NEA/IEA)

PERSPECTIVES DE L'ÉNERGIE NUCLÉAIRE

Perspectives de l'Énergie Nucléaire jusqu'en 2000
(Rapport conjoint AEN/AIE)

£7.00 US$14.00 F70,00

SCIENTIFIC AND TECHNICAL PUBLICATIONS

PUBLICATIONS SCIENTIFIQUES ET TECHNIQUES

NUCLEAR FUEL CYCLE

LE CYCLE DU COMBUSTIBLE NUCLÉAIRE

World Uranium Potential – An International Evaluation (1978)

Potentiel mondial en uranium – Une évaluation internationale (1978)

£7.80 US$16.00 F64.00

Uranium – Ressources, Production and Demand (1982)

Uranium – ressources, production et demande (1982)

£9.90 US$22.00 F99,00

Nuclear Energy and Its Fuel Cycle: Prospects to 2025

L'énergie nucléaire et son cycle de combustible : perspectives jusqu'en 2025

£11.00 US$24.00 F110,00

Dry Storage of Spent Fuel Elements (Proceedings of an NEA Specialist Workshop, Madrid, 1982) [in preparation].

Stockage à sec des éléments combustibles irradiés (Compte rendu d'une réunion de spécialistes de l'AEN, Madrid, 1982). [en préparation].

£ US$ F

Uranium Exploration Methods – Review of the NEA/IAEA R & D Programme (Proceedings of the Paris Symposium, 1982) [in preparation].

Les méthodes de prospection de l'uranium – Examen du programme AEN/AIEA de R & D (Compte rendu du symposium de Paris, 1982) [en préparation].

£ 24.00 US$ 48.00 F 240,00

• • •

SCIENTIFIC INFORMATION

INFORMATION SCIENTIFIQUE

Calculation of 3-Dimensional Rating Distributions in Operating Reactors (Proceedings of the Paris Specialists' Meeting, 1979)

Calcul des distributions tridimensionnelles de densité de puissance dans les réacteurs en cours d'exploitation (Compte rendu de la Réunion de spécialistes de Paris, 1979)

£9.60 US$21.50 F86.00

Nuclear Data and Benchmarks for Reactor Shielding (Proceedings of a Specialists' Meeting, Paris, 1980)

Données nucléaires et expériences repères en matière de protection des réacteurs (Compte rendu d'une réunion de spécialistes, Paris, 1980)

£9.60 US$24.00 F96,00

• • •

RADIATION PROTECTION

RADIOPROTECTION

Iodine-129
(Proceedings of an NEA Specialist Meeting, Paris, 1977)

Iode-129
(Compte rendu d'une réunion de spécialistes de l'AEN, Paris, 1977)

£3.40 US$7.00 F28,00

Recommendations for Ionization Chamber Smoke Detectors in Implementation of Radiation Protection Standards (1977)

Recommandations relatives aux détecteurs de fumée à chambre d'ionisation en application des normes de radioprotection (1977)

Free on request – Gratuit sur demande

Radon Monitoring
(Proceedings of the NEA Specialist Meeting, Paris, 1978)

Surveillance du radon
(Compte rendu d'une réunion de spécialistes de l'AEN, Paris, 1978)

£8.00 US$16.50 F66,00

Management, Stabilisation and Environmental Impact of Uranium Mill Tailings
(Proceedings of the Albuquerque Seminar, United States, 1978)

Gestion, stabilisation et incidence sur l'environnement des résidus de traitement de l'uranium
(Compte rendu du Séminaire d'Albuquerque, États-Unis, 1978)

£9.80 US$20.00 F80,00

Exposure to Radiation from the Natural Radioactivity in Building Materials
(Report by an NEA Group of Experts, 1979)

Exposition aux rayonnements due à la radioactivité naturelle des matériaux de construction
(Rapport établi par un Groupe d'experts de l'AEN, 1979)

Free on request – Gratuit sur demande

Marine Radioecology
(Proceedings of the Tokyo Seminar, 1979)

Radioécologie marine
(Compte rendu du Colloque de Tokyo, 1979)

£9.60 US$21.50 F86.00

Radiological Significance and Management of Tritium, Carbon-14, Krypton-85 and Iodine-129 arising from the Nuclear Fuel Cycle
(Report by an NEA Group of Experts, 1980)

Importance radiologique et gestion des radionucléides : tritium, carbone-14, krypton-85 et iode-129, produits au cours du cycle du combustible nucléaire
(Rapport établi par un Groupe d'experts de l'AEN, 1980)

£8.40 US$19.00 F76,00

The Environmental and Biological Behaviour of Plutonium and Some Other Transuranium Elements (Report by an NEA Group of Experts, 1981)

Le comportement mésologique et biologique du plutonium et de certains autres éléments transuraniens (Rapport établi par un Groupe d'experts de l'AEN, 1981)

£4.60 US$10.00 F46,00

Uranium Mill Tailings Management
(Proceedings of two Workshops)
(in preparation)

La gestion des résidus de traitement de l'uranium
(Compte rendu de deux réunions de travail) (en préparation)

£7.20 US$16.00 F72,00

RADIOACTIVE WASTE MANAGEMENT — GESTION DES DÉCHETS RADIOACTIFS

Objectives, Concepts and Strategies for the Management of Radioactive Waste Arising from Nuclear Power Programmes (Report by an NEA Group of Experts, 1977)

Objectifs, concepts et stratégies en matière de gestion des déchets radioactifs résultant des programmes nucléaires de puissance (Rapport établi par un Groupe d'experts de l'AEN, 1977)

£8.50 US$17.50 F70,00

Treatment, Conditioning and Storage of Solid Alpha-Bearing Waste and Cladding Hulls (Proceedings of the NEA/IAEA Technical Seminar, Paris, 1977)

Traitement, conditionnement et stockage des déchets solides alpha et des coques de dégainage (Compte rendu du Séminaire technique AEN/AIEA, Paris, 1977)

£7.30 US$15.00 F60,00

Storage of Spent Fuel Elements (Proceedings of the Madrid Seminar, 1978)

Stockage des éléments combustibles irradiés (Compte rendu du Séminaire de Madrid, 1978)

£7.30 US$15.00 F60,00

In Situ Heating Experiments in Geological Formations (Proceedings of the Ludvika Seminar, Sweden, 1978)

Expériences de dégagement de chaleur in situ dans les formations géologiques (Compte rendu du Séminaire de Ludvika, Suède, 1978)

£8.00 US$16.50 F66,00

Migration of Long-lived Radionuclides in the Geosphere (Proceedings of the Brussels Workshop, 1979)

Migration des radionucléides à vie longue dans la géosphère (Compte rendu de la réunion de travail de Bruxelles, 1979)

£8.30 US$17.00 F68,00

Low-Flow, Low-Permeability Measurements in Largely Impermeable Rocks (Proceedings of the Paris Workshop, 1979)

Mesures des faibles écoulements et des faibles perméabilités dans des roches relativement imperméables (Compte rendu de la réunion de travail de Paris, 1979)

£7.80 US$16.00 F64,00

On-Site Management of Power Reactor Wastes (Proceedings of the Zurich Symposium, 1979)

Gestion des déchets en provenance des réacteurs de puissance sur le site de la centrale (Compte rendu du Colloque de Zurich, 1979)

£11.00 US$22.50 F90,00

Recommended Operational Procedures for Sea Dumping of Radioactive Waste (1979)

Recommandations relatives aux procédures d'exécution des opérations d'immersion de déchets radioactifs en mer (1979)

Free on request — Gratuit sur demande

Guidelines for Sea Dumping Packages of Radioactive Waste (Revised version, 1979)

Guide relatif aux conteneurs de déchets radioactifs destinés au rejet en mer (Version révisée, 1979)

Free on request — Gratuit sur demande

Use of Argillaceous Materials for the Isolation of Radioactive Waste (Proceedings of the Paris Workshop, 1979)

Utilisation des matériaux argileux pour l'isolement des déchets radioactifs (Compte rendu de la Réunion de travail de Paris, 1979)

£7.60 US$17.00 F68,00

Review of the Continued Suitability of the Dumping Site for Radioactive Waste in the North-East Atlantic (1980)

Réévaluation de la validité du site d'immersion de déchets radioactifs dans la région nord-est de l'Atlantique (1980)

Free on request — Gratuit sur demande

Decommissioning Requirements in the Design of Nuclear Facilities (Proceedings of the NEA Specialist Meeting, Paris, 1980)

Déclassement des installations nucléaires : exigences à prendre en compte au stade de la conception (Compte rendu d'une réunion de spécialistes de l'AEN, Paris, 1980)

£7.80 US$17.50 F70,00

Borehole and Shaft Plugging (Proceedings of the Columbus Workshop, United States, 1980)

Colmatage des forages et des puits (Compte rendu de la réunion de travail de Columbus, États-Unis, 1980)

£12.00 US$30.00 F120,00

Radionucleide Release Scenarios for Geologic Repositories (Proceedings of the Paris Workshop, 1980)

Scénarios de libération des radionucléides à partir de dépôts situés dans les formations géologiques (Compte rendu de la réunion de travail de Paris, 1980)

£6.00 US$15.00 F60,00

Research and Environmental Surveillance Programme Related to Sea Disposal of Radioactive Waste (1981)

Programme de recherches et de surveillance du milieu lié à l'immersion de déchets radioactifs en mer (1981)

Free on request — Gratuit sur demande

Cutting Techniques as related to Decommissioning of Nuclear Facilities (Report by an NEA Group of Experts, 1981)

Techniques de découpe utilisées au cours du déclassement d'installations nucléaires (Rapport établi par un Groupe d'experts de l'AEN, 1981)

£3.00 US$7.50 F30.00

Decontamination Methods as related to Decommissioning of Nuclear Facilities (Report by an NEA Group of Experts, 1981)

Méthodes de décontamination relatives au déclassement des installations nucléaires (Rapport établi par un Groupe d'experts de l'AEN, 1981)

£2.80 US$7.00 F28,00

Siting of Radioactive Waste Repositories in Geological Formations (Proceedings of the Paris Workshop, 1981)

Choix des sites des dépôts de déchets radioactifs dans les formations géologiques (Compte rendu d'une réunion de travail de Paris, 1981)

£6.80 US$15.00 F68,00

Near-Field Phenomena in Geologic Repositories for Radioactive Waste (Proceedings of the Seattle Workshop, United States, 1981)

Phénomènes en champ proche des dépôts de déchets radioactifs en formations géologiques (Compte rendu de la réunion de travail de Seattle, Etats-Unis, 1981)

£11.00 $24.50 F110,00

Disposal of Radioactive Waste – An Overview of the Principles Involved, 1982

Évacuation des déchets radioactifs – un aperçu des principes en vigueur, 1982

Free on request – Gratuit sur demande

Geological Disposal of Radioactive Waste – Research in the OECD Area (1982)

Évacuation des déchets radioactifs dans les formations géologiques – Recherches effectuées dans les pays de l'OCDE (1982).

Free on request – Gratuit sur demande

• • •

SAFETY

SÛRETÉ

Safety of Nuclear Ships (Proceedings of the Hamburg Symposium, 1977)

Sûreté des navires nucléaires (Compte rendu du Symposium de Hambourg, 1977)

£17.00 US$35.00 F140,00

Nuclear Aerosols in Reactor Safety (A State-of-the-Art Report by a Group of Experts, 1979)

Les aérosols nucléaires dans la sûreté des réacteurs (Rapport sur l'état des connaissances établi par un Groupe d'Experts, 1979)

£8.30 US$18.75 F75,00

Plate Inspection Programme (Report from the Plate Inspection Steering Committee – PISC – on the Ultrasonic Examination of Three Test Plates), 1980

Programme d'inspection des tôles (Rapport du Comité de Direction sur l'inspection des tôles – PISC – sur l'examen par ultrasons de trois tôles d'essai au moyen de la procédure «PISC» basée sur le code ASME XI), 1980

£3.30 US$7.50 F30.00

Reference Seismic Ground Motions in Nuclear Safety Assessments (A State-of-the-Art Report by a Group of Experts, 1980)

Les mouvements sismiques de référence du sol dans l'évaluation de la sûreté des installations nucléaires (Rapport sur l'état des connaissances établi par un Groupe d'experts, 1980)

£7.00 US$16.00 F64,00

Nuclear Safety Research in the OECD Area. The Response to the Three Mile Island Accident (1980)

Les recherches en matière de sûreté nucléaire dans les pays de l'OCDE. L'adaptation des programmes à la suite de l'accident de Three Mile Island (1980)

£3.20 US$8.00 F32,00

Safety Aspects of Fuel Behaviour in Off-Normal and Accident Conditions (Proceedings of the Specialist Meeting, Espoo, Finland, 1980)

Considérations de sûreté relatives au comportement du combustible dans des conditions anormales et accidentelles (Compte rendu de la réunion de spécialistes, Espoo, Finlande, 1980)

£12.60 $28.00 F126,00

Safety of the Nuclear Fuel Cycle (A State-of-the-Art Report by a Group of Experts, 1981)

Sûreté du Cycle du Combustible Nucléaire (Rapport sur l'état des connaissances établi par un Groupe d'Experts, 1981)

£6.60 $16.50 F66,00

LEGAL PUBLICATIONS — PUBLICATIONS JURIDIQUES

Convention on Third Party Liability in the Field of Nuclear Energy — incorporating the provisions of Additional Protocol of January 1964

Convention sur la responsabilité civile dans le domaine de l'énergie nucléaire — Texte incluant les dispositions du Protocole additionnel de janvier 1964

Free on request — Gratuit sur demande

Nuclear Legislation, Analytical Study: "Nuclear Third Party Liability" (revised version, 1976)

Législations nucléaires, étude analytique: "Responsabilité civile nucléaire" (version révisée, 1976)

£6.00 US$12.50 F50,00

Nuclear Legislation, Analytical Study: "Regulations governing the Transport of Radioactive Materials" (1980)

Législations nucléaires, étude analytique : "Réglementation relative au transport des matières radioactives" (1980)

£8.40 US$21.00 F84,00

Nuclear Law Bulletin
(Annual Subscription — two issues and supplements)

Bulletin de Droit Nucléaire
(Abonnement annuel — deux numéros et suppléments)

£6.00 $13.00 F60,00

Index of the first twenty five issues of the Nuclear Law Bulletin

Index des vingt-cinq premiers numéros du Bulletin de Droit Nucléaire

Description of Licensing Systems and Inspection of Nuclear Installation (1980)

Description du régime d'autorisation et d'inspection des installations nucléaires (1980)

£7.60 US$19.00 F76,00

NEA Statute

Statuts de l'AEN

Free on request — Gratuit sur demande

• • •

OECD SALES AGENTS
DÉPOSITAIRES DES PUBLICATIONS DE L'OCDE

ARGENTINA – ARGENTINE
Carlos Hirsch S.R.L., Florida 165, 4° Piso (Galería Guemes)
1333 BUENOS AIRES, Tel. 33.1787.2391 y 30.7122

AUSTRALIA – AUSTRALIE
Australia and New Zealand Book Company Pty, Ltd.,
10 Aquatic Drive, Frenchs Forest, N.S.W. 2086
P.O. Box 459, BROOKVALE, N.S.W. 2100

AUSTRIA – AUTRICHE
OECD Publications and Information Center
4 Simrockstrasse 5300 BONN. Tel. (0228) 21.60.45
Local Agent/Agent local :
Gerold and Co., Graben 31, WIEN 1. Tel. 52.22.35

BELGIUM – BELGIQUE
LCLS
35, avenue de Stalingrad, 1000 BRUXELLES. Tel. 02.512.89.74

BRAZIL – BRÉSIL
Mestre Jou S.A., Rua Guaipa 518,
Caixa Postal 24090, 05089 SAO PAULO 10. Tel. 261.1920
Rua Senador Dantas 19 s/205-6, RIO DE JANEIRO GB.
Tel. 232.07.32

CANADA
Renouf Publishing Company Limited,
2182 St. Catherine Street West,
MONTRÉAL, Que. H3H 1M7. Tel. (514)937.3519
OTTAWA, Ont. K1P 5A6, 61 Sparks Street

DENMARK – DANEMARK
Munksgaard Export and Subscription Service
35, Nørre Søgade
DK 1370 KØBENHAVN K. Tel. +45.1.12.85.70

FINLAND – FINLANDE
Akateeminen Kirjakauppa
Keskuskatu 1, 00100 HELSINKI 10. Tel. 65.11.22

FRANCE
Bureau des Publications de l'OCDE,
2 rue André-Pascal, 75775 PARIS CEDEX 16. Tel. (1) 524.81.67
Principal correspondant :
13602 AIX-EN-PROVENCE : Librairie de l'Université.
Tel. 26.18.08

GERMANY – ALLEMAGNE
OECD Publications and Information Center
4 Simrockstrasse 5300 BONN Tel. (0228) 21.60.45

GREECE – GRÈCE
Librairie Kauffmann, 28 rue du Stade,
ATHÈNES 132. Tel. 322.21.60

HONG-KONG
Government Information Services,
Publications/Sales Section, Baskerville House,
2/F., 22 Ice House Street

ICELAND – ISLANDE
Snaebjörn Jönsson and Co., h.f.,
Hafnarstraeti 4 and 9, P.O.B. 1131, REYKJAVIK.
Tel. 13133/14281/11936

INDIA – INDE
Oxford Book and Stationery Co. :
NEW DELHI-1, Scindia House. Tel. 45896
CALCUTTA 700016, 17 Park Street. Tel. 240832

INDONESIA – INDONÉSIE
PDIN-LIPI, P.O. Box 3065/JKT., JAKARTA, Tel. 583467

IRELAND – IRLANDE
TDC Publishers – Library Suppliers
12 North Frederick Street, DUBLIN 1 Tel. 744835-749677

ITALY – ITALIE
Libreria Commissionaria Sansoni :
Via Lamarmora 45, 50121 FIRENZE. Tel. 579751
Via Bartolini 29, 20155 MILANO. Tel. 365083
Sub-depositari :
Editrice e Libreria Herder,
Piazza Montecitorio 120, 00 186 ROMA. Tel. 6794628
Libreria Hoepli, Via Hoepli 5, 20121 MILANO. Tel. 865446
Libreria Lattes, Via Garibaldi 3, 10122 TORINO. Tel. 519274
La diffusione delle edizioni OCSE è inoltre assicurata dalle migliori librerie nelle città più importanti.

JAPAN – JAPON
OECD Publications and Information Center,
Landic Akasaka Bldg., 2-3-4 Akasaka,
Minato-ku, TOKYO 107 Tel. 586.2016

KOREA – CORÉE
Pan Korea Book Corporation,
P.O. Box n° 101 Kwangwhamun, SÉOUL. Tel. 72.7369

LEBANON – LIBAN
Documenta Scientifica/Redico,
Edison Building, Bliss Street, P.O. Box 5641, BEIRUT.
Tel. 354429 – 344425

MALAYSIA – MALAISIE
and/et SINGAPORE - SINGAPOUR
University of Malaysia Co-operative Bookshop Ltd.
P.O. Box 1127, Jalan Pantai Baru
KUALA LUMPUR. Tel. 51425, 54058, 54361

THE NETHERLANDS – PAYS-BAS
Staatsuitgeverij
Verzendboekhandel Chr. Plantijnstraat 1
Postbus 20014
2500 EA S-GRAVENHAGE. Tel. nr. 070.789911
Voor bestellingen: Tel. 070.789208

NEW ZEALAND – NOUVELLE-ZÉLANDE
Publications Section,
Government Printing Office Bookshops:
AUCKLAND: Retail Bookshop: 25 Rutland Street,
Mail Orders: 85 Beach Road, Private Bag C.P.O.
HAMILTON: Retail Ward Street,
Mail Orders, P.O. Box 857
WELLINGTON: Retail: Mulgrave Street (Head Office),
Cubacade World Trade Centre
Mail Orders: Private Bag
CHRISTCHURCH: Retail: 159 Hereford Street,
Mail Orders: Private Bag
DUNEDIN: Retail: Princes Street
Mail Order: P.O. Box 1104

NORWAY – NORVÈGE
J.G. TANUM A/S Karl Johansgate 43
P.O. Box 1177 Sentrum OSLO 1. Tel. (02) 80.12.60

PAKISTAN
Mirza Book Agency, 65 Shahrah Quaid-E-Azam, LAHORE 3.
Tel. 66839

PHILIPPINES
National Book Store, Inc.
Library Services Division, P.O. Box 1934, MANILA.
Tel. Nos. 49.43.06 to 09, 40.53.45, 49.45.12

PORTUGAL
Livraria Portugal, Rua do Carmo 70-74,
1117 LISBOA CODEX. Tel. 360582/3

SPAIN – ESPAGNE
Mundi-Prensa Libros, S.A.
Castelló 37, Apartado 1223, MADRID-1. Tel. 275.46.55
Libreria Bosch, Ronda Universidad 11, BARCELONA 7.
Tel. 317.53.08, 317.53.58

SWEDEN – SUÈDE
AB CE Fritzes Kungl Hovbokhandel,
Box 16 356, S 103 27 STH, Regeringsgatan 12,
DS STOCKHOLM. Tel. 08/23.89.00

SWITZERLAND – SUISSE
OECD Publications and Information Center
4 Simrockstrasse 5300 BONN. Tel. (0228) 21.60.45
Local Agents/Agents locaux
Librairie Payot, 6 rue Grenus, 1211 GENÈVE 11. Tel. 022.31.89.50
Freihofer A.G., Weinbergstr. 109, CH-8006 ZÜRICH.
Tel. 01.3634282

TAIWAN – FORMOSE
Good Faith Worldwide Int'l Co., Ltd.
9th floor, No. 118, Sec. 2
Chung Hsiao E. Road
TAIPEI. Tel. 391.7396/391.7397

THAILAND – THAILANDE
Suksit Siam Co., Ltd., 1715 Rama IV Rd,
Samyan, BANGKOK 5. Tel. 2511630

TURKEY – TURQUIE
Kültur Yayinlari Is-Türk Ltd. Sti.
Atatürk Bulvari No : 77/B
KIZILAY/ANKARA. Tel. 17 02 66
Dolmabahce Cad. No : 29
BESIKTAS/ISTANBUL. Tel. 60 71 88

UNITED KINGDOM – ROYAUME-UNI
H.M. Stationery Office, P.O.B. 569,
LONDON SE1 9NH. Tel. 01.928.6977, Ext. 410 or
49 High Holborn, LONDON WC1V 6 HB (personal callers)
Branches at: EDINBURGH, BIRMINGHAM, BRISTOL,
MANCHESTER, CARDIFF, BELFAST.

UNITED STATES OF AMERICA – ÉTATS-UNIS
OECD Publications and Information Center, Suite 1207,
1750 Pennsylvania Ave., N.W. WASHINGTON, D.C.20006 – 4582
Tel. (202) 724.1857

VENEZUELA
Libreria del Este, Avda. F. Miranda 52, Edificio Galipan,
CARACAS 106. Tel. 32.23.01/33.26.04/33.24.73

YUGOSLAVIA – YOUGOSLAVIE
Jugoslovenska Knjiga, Terazije 27, P.O.B. 36, BEOGRAD.
Tel. 621.992

Les commandes provenant de pays où l'OCDE n'a pas encore désigné de dépositaire peuvent être adressées à :
OCDE, Bureau des Publications, 2, rue André-Pascal, 75775 PARIS CEDEX 16.

Orders and inquiries from countries where sales agents have not yet been appointed may be sent to:
OECD, Publications Office, 2 rue André-Pascal, 75775 PARIS CEDEX 16.

OECD PUBLICATIONS, 2, rue André-Pascal, 75775 PARIS CEDEX 16 - No. 42263 1982
PRINTED IN FRANCE
(66 82 05 1) ISBN 92-64-12366-0